Power Systems

Electrical power has been the technological foundation of industrial societies for many years. Although the systems designed to provide and apply electrical energy have reached a high degree of maturity, unforeseen problems are constantly encountered, necessitating the design of more efficient and reliable systems based on novel technologies. The book series Power Systems is aimed at providing detailed, accurate and sound technical information about these new developments in electrical power engineering. It includes topics on power generation, storage and transmission as well as electrical machines. The monographs and advanced textbooks in this series address researchers, lecturers, industrial engineers and senior students in electrical engineering.

Power Systems is indexed in Scopus

Vasily Ya. Ushakov · Ikromjon U. Rakhmonov ·
Alisher B. Askarov · Dmitriy S. Nikitin

Digitalization of Electrical Power Engineering

Scientific and Technical Fundamentals and Achieved Advantages

Springer

Vasily Ya. Ushakov
National Research Tomsk Polytechnic University
Tomsk, Russia

Alisher B. Askarov
National Research Tomsk Polytechnic University
Tomsk, Russia

Ikromjon U. Rakhmonov
Tashkent State Technical University
Tashkent, Uzbekistan

Dmitriy S. Nikitin
National Research Tomsk Polytechnic University
Tomsk, Russia

ISSN 1612-1287 ISSN 1860-4676 (electronic)
Power Systems
ISBN 978-3-031-95704-8 ISBN 978-3-031-95705-5 (eBook)
https://doi.org/10.1007/978-3-031-95705-5

This Springer imprint is published by the registered company Springer Nature Switzerland AG
The registered company address is: Gewerbestrasse 11, 6330 Cham, Switzerland

Preface

The primary goal of the EPE, as a fundament of the economy in all developed countries, is to ensure the reliable supply of high-quality electricity at affordable prices while minimizing negative environmental impacts. Throughout the past century, the EPE has continued to make a decisive contribution to the welfare and overall progress of humanity. The growth in electricity consumption worldwide has been phenomenal. From the beginning of the twentieth century to the early 1970s, electricity sales increased hundreds of times, and this trend continues today [1]. Many experts believe that the growing demand for electricity can only be met through revolutionary changes in the EPE.

The experience of the leading industrialized countries with advanced digital technology markets (the USA, European Union countries, China, Russia, and India) demonstrates that maintaining and improving human well-being in the face of increasing challenges requires the intellectualization of all sectors of the economy based on digitalization, known as the "digital transition", which includes (and perhaps primarily concerns) the EPE.

The purpose of this textbook is to provide readers with practical knowledge of the fundamentals of the "Digital Transition" concept, tools for designing and managing the operation of smart grids and substations, and critical issues in modernizing the EPE. It summarizes materials from numerous monographs, scientific articles, and conference reports discussing these issues.

In these materials, two terms are often used interchangeably: "Smart (or intelligent) electrical grid" and "Intelligent (or digital) electrical power industry". To simplify the reader's task of relating the content of this textbook to materials from other publications, the authors of this textbook frequently do the same. However, it is important to note that, while there are differences between the two terms, they are not fundamental.

To transform a traditional electrical grid into a smart grid, it will be enhanced with intelligent (digital) sensors and meters, digital control tools, and analytical instruments that enable automation, monitoring, and control of energy transmission from the power plant to the consumer. Energy companies will be able to increase the efficiency of grid operations, prevent power outages, or speed up the restoration of electricity supply. Additionally, they will provide customers with the ability to manage energy consumption down to the level of individual devices connected to the grid. Moreover, rising electricity prices and technological advancements are changing the collective mindset of consumers. As a result, many consumers are transforming from passive "bill-payers" into well-informed customers who, under certain conditions, become participants in electricity production and supply excess energy back to the grid.

The use of the term "intelligent electrical power industry" emphasizes the provision of necessary information and management tools by energy companies to consumers, enabling them to change their behavior models and thereby increase the efficiency of electricity use. This marks a gradual transition from smart grids to intelligent energy systems, where grids are undoubtedly a critical component but not the only one. The term "intelligent energy" is therefore a broader concept that encompasses all stages of the technological (electrical power) chain—from electricity generation to its consumption.

Given the limited scope of the textbook, certain issues are discussed very briefly; in fact, they are merely meant to make the reader aware of their existence, with the expectation that they will refer to primary sources if necessary.

The textbook contains an Introduction, 11 chapters, and Conclusion. Each chapter begins with a brief annotation and concludes with a set of control questions and tasks and a list of references.

Feedback and suggestions are welcomed and can be sent to the following email addresses: vyush@tpu.ru and ilider1987@yandex.com.

Tomsk, Russia Vasily Ya. Ushakov
Tashkent, Uzbekistan Ikromjon U. Rakhmonov
Tomsk, Russia Alisher B. Askarov
Tomsk, Russia Dmitriy S. Nikitin

Reference

1. Ushakov V (2018) Electrical power engineering: current state. Springer, Problems and Perspectives, p 258

Competing Interests The authors have no competing interests to declare that are relevant to the content of this manuscript.

About This Book

The inevitable transformation of the electrical power engineering (EPE) is moving towards digitalization, which has taken the form of the concept of "Smart (Intelligent) grids" (sometimes called "Digital Transition"). This has required coordinated work not only from the EPE but also from other stakeholders (government agencies, private firms, utilities, scientific and educational institutions, etc.) to solve numerous problems that arise during the transition to this advanced technology. It is obvious that the development of new components of energy systems, such as smart meters, monitoring and geolocation equipment, new disaster recovery systems, and energy storage systems, should be supplemented by the creation of the necessary communication infrastructure, modernization of existing network components, implementation of relevant regulations, and training of highly qualified personnel.

The textbook covers all the main tasks and ways to solve them for the implementation of this technology. The authors hope that the tasks and means necessary for the implementation of the concept of "Smart Grids" are considered in it in a volume sufficient for students, postgraduates, and young specialists studying or working in the field of "Electric Power Engineering" and related areas.

Contents

1 **Intellectualization (Digitalization)—The Main Direction of Reforming the Electrical Power Engineering** ... 1
1.1 Current Problems in the Electrical Power Engineering ... 1
1.2 The Concept of Digitalization and Its Main Objectives ... 3
1.3 Advantages of Smart Grids Compared to Traditional Grids ... 7
1.4 Human Resource Development for the Intellectualization of the Electrical Power Engineering ... 11
1.5 Requirements for University Graduates for Effective Participation in the "Digital Transition" in the Electrical Power Engineering ... 12
References ... 14

2 **Stakeholder Motives and Participation in the Digital Transition** ... 15
2.1 Roles and Functions of Stakeholders ... 15
2.2 Concentration of Efforts by Employees of Scientific Institutions, University Teachers and Engineers of Energy Companies on the Creation of Smart Grids ... 18
2.2.1 Industry Laboratories, Research Institutions, and Universities ... 18
2.2.2 Technology Companies, Manufacturers, and Suppliers ... 19
References ... 19

3 **The Smart Electric Grid** ... 21
3.1 Key Principles ... 21
3.2 Key Components of the System Enabling the Smart Grid and Their Functions ... 22
3.2.1 Energy Distribution Control Systems ... 22
3.2.2 Demand Management Systems ... 23
3.2.3 New Electrical Power Consumption Metering Systems ... 23

3.3 Traditional High-Voltage Elements of the Electric Grid 24
3.3.1 Overhead Transmission and Cable Lines 24
3.3.2 Energy Storage Systems 24
References 26

4 Testing and Monitoring Systems for the Electrical Grid 27
4.1 Sensors 27
4.2 Testing 29
4.3 Monitoring 32
4.3.1 Monitoring Software Development 32
4.3.2 Non-Contact Fault Location on Transmission Lines Using a Magnetic Sensor 33
4.3.3 Monitoring the Operational State of High-Voltage Transmission Lines with a Magnetoresistive Sensor 34
4.3.4 Current Monitoring in the Grid Using a MR Sensors 34
4.4 Supervisory Control and Data Acquisition Systems (SCADA) 35
4.4.1 Functional and Communication Requirements for SCADA Systems 35
4.4.2 Functions of SCADA Systems 39
4.4.3 Components of a SCADA System 40
References 42

5 Digital Substations as a Key Element of Smart Grids 45
5.1 Main Provisions 45
5.2 Structure of a Digital Substation 48
5.3 Elements of Digital Substations 51
5.3.1 Electronic Current and Voltage Transformers 52
5.3.2 Station and Process Buses: Three Levels of Communication 54
5.4 Integrated Information Platform and Advanced Applications 56
5.5 Intelligent Condition Monitoring of Primary Equipment 59
5.5.1 Point Measurement of Transient Magnetic Fields in Substations Using a Magnetoresistive Sensor 61
5.5.2 Online Monitoring of Power Transformers, Gas-Insulated Switchgears and Surge Arresters 61
5.6 Digital Substation Protection 63
References 65

6 Digitalization of Electric Power Grid Within IEC 61850 Standard 67
6.1 Content and Scope of the IEC 61850 Standard 67
6.1.1 Scope of the IEC 61850 Standard 67
6.1.2 Overview of the IEC 61850 Standard 68

6.2 IEC 61850-Based Information Model 74
6.2.1 Main Elements of the Information Model 75
6.2.2 Logical Node and Logical Device 78
6.2.3 Data Objects and Attributes 79
6.2.4 Data Modelling Procedure 86
References 90

7 IEC 61850 Requirements for Digital Substation Design 91
7.1 SCL Configuration Language 91
7.1.1 Types of SCL Files 91
7.1.2 Classification of Software Tools 97
7.2 Digital Substation Design 100
7.2.1 Design Process According to IEC 61850 Standard 100
7.2.2 Purpose and Structure of SSD, ICD and SCD Files 102
7.2.3 File Validation 108
References 113

8 IEC 61850 Digital Communication Protocols 115
8.1 Data Communication Services and Protocols 115
8.1.1 Purpose of Abstract Communication Models and Services in IEC 61850 115
8.1.2 Communication Protocol Mapping 117
8.2 Application of Manufacturing Message Specification (MMS) Protocol 119
8.2.1 Reports and Event Logs 119
8.2.2 Operational Control 125
8.2.3 Changing Setpoints and Setpoint Groups 128
8.3 Generic Object Oriented Substation Event (GOOSE) Protocol 131
8.3.1 Abstract Model, GOOSE Message Transmission Block, and Communication Services 132
8.3.2 Configuration of GOOSE Message Transmission in SCL Language 137
8.3.3 Mapping the GOOSE Abstract Model to a Specific Communication Protocol 140
8.4 Sampled Measured Values (SV) Protocol 145
8.4.1 Abstract Model, Sampled Values Control Block, and Communication Services 145
8.4.2 Mapping the Sampled Values Abstract Model to a Specific Communication Protocol 149
8.4.3 Configuration of Sampled Values Transmission in SCL Language 151
References 154

9 Information Security of the Digital Substation 157
9.1 Key Provisions 157
9.2 Cyberattacks and Cybersecurity Threats at Substations 158
9.2.1 Intentional Electromagnetic Interference 158
9.2.2 Deployment and Maintenance Platform Vulnerabilities 160
9.2.3 Firewall Configuration Errors 160
9.2.4 Online Password Cracking 160
9.2.5 Inadequate Access Control Measures 160
9.2.6 Network Traffic Analysis and Manipulation 161
9.3 Ensuring Cybersecurity 161
9.3.1 Practical Cybersecurity Solutions 161
9.3.2 Cyberattack Countermeasures by Substation Automation Engineers 165
9.3.3 Cybersecurity Testing Platform for IEC 61850 Digital Substations 166
9.3.4 Cyber Vulnerability Studies on Digital Substations 168
References 170

10 Application of Artificial Intelligence (AI) and Deep/Machine Learning (DL/ML) in Electrical Power Engineering 171
10.1 General Use of AI and DL/ML to Improve Energy Efficiency 171
10.2 Need for AI and DL/ML in Smart Electric Grid 174
10.3 Types of AI and DL/ML Techniques for Power System 176
10.4 Role of AI and DL/ML for Power System Stability, Control, Security and Protection 180
10.5 Today's Challenges and Future Scope of AI in Power System ... 183
References 185

11 Impact of the Application of Various Digitalization Tools on the Energy Efficiency of Industrial Enterprises 187
11.1 Internet of Things (IoT) for Resource Management 187
11.2 Analysis of Energy Consumption Using Big Data 188
11.3 Cloud Technologies for Making Efficiency Gains 189
11.4 Models and Digital Twins of Power Systems and Industrial Processes 191
References 193

Conclusion 195

Chapter 1
Intellectualization (Digitalization)—The Main Direction of Reforming the Electrical Power Engineering

1.1 Current Problems in the Electrical Power Engineering

The numerous problems that global energy systems face in achieving the goal stated in the Introduction can be grouped into four categories:

- depletion of non-renewable (mineral) energy resources ("energy hunger"),
- negative environmental impact from energy facilities ("ecological crisis"/("ecological infarct"),
- significant wear and tear of power infrastructure,
- social and geopolitical threats associated with meeting the growing electricity needs of humanity.

The first group of problems is driven by the rapid growth in per capita electricity consumption and the increasing global population. Today, this is particularly relevant for countries that do not have sufficient domestic mineral energy resources and cannot afford to import them due to financial constraints [1, 2]. Many experts assert that the key issue in this century will not be the shortage of energy resources but the lack of investments due to rising costs.

Many scholars studying the issue of sustainable development for human society argue that **the second group of problems**—namely, the harmful environmental impact of industries, particularly the fuel and energy complex (FEC)—poses no less of a threat [2, 3]. However, this view has its opponents, who point to the excessive politicization involved in assessing this danger.

The third group of problems, affecting many countries including Russia, is due to the significant wear and tear of key elements of power systems. According to various estimates, about 60–70% of the core assets of the power grid complex have exceeded their service life. With the sharp increase in electricity consumption by both industrial facilities and the residential sector, dispatch management using existing automated process control systems (APCS) and supervisory control and data acquisition (SCADA) systems struggles to cope with emergencies, leading to

V. Y. Ushakov et al., *Digitalization of Electrical Power Engineering*, Power Systems,
https://doi.org/10.1007/978-3-031-95705-5_1

failures and cascading power outages across sections of the power grid, accompanied by significant losses.

The energy system is considered one of the most reliable technical systems. For example, in the United States, the reliability of even the traditional power system is 99.97%. Nevertheless, it still allows for outages and interruptions in power supply, which cost Americans at least $150 billion each year—around $500 per person [4].

It is evident that replacing the technically and morally obsolete, and now economically inefficient, infrastructure of power systems is an objective necessity for both the industry and society [5].

The relevance of **the fourth group of problems**—social and geopolitical threats—is well emphasized by journalists: "Today's wars not only smell of blood but also of oil" (quote from [6]). The reality of this issue is quickly and easily confirmed when marking on a world map the major locations of mineral energy resources and the sites of political and military conflicts that have occurred in the last 3–4 decades.

The methods for addressing the challenges of the first two groups are based on the adoption of new, more efficient technologies for the generation, transmission, and distribution of electricity, as well as on improving energy consumption efficiency. The latter includes encouraging every consumer to take a more careful approach to energy saving.

In principle, two scenarios for the development of the global electric power industry are possible:

1. Maintaining power facilities in working condition, limited, if necessary, to restoring their previous parameters using the already mastered component base and previous circuit solutions. This approach, like any "patching of holes", has its advantages—relatively low required investments and short implementation timelines due to the use of equipment long produced by industry and familiar to the personnel of power companies. This path can be considered a fallback option in the event of economic crises and depressions. For example, during the crisis of the 1990s, the Russian energy industry largely survived in this way.
2. The transforming all key components of the electric power system (EPS) into smart grids, smart substations, etc. In terms of depth and scale, this transformation can be considered revolutionary, but in terms of pace, it is rather evolutionary in nature, since it is capable of significantly improving the operation of the existing EPS before its complete transformation.

Such efficient technologies and their components, including intelligent devices, communication and information systems, as well as operational software, play a crucial role in creating smart grids, which have become the main direction for modernizing existing power grids and constructing new ones worldwide. Thanks to the application of the latest advancements in metrology, communications, information technology, artificial intelligence (AI), energy management, power systems are undergoing a fundamental transformation to achieve the following key objectives:

1. Reinvesting in the infrastructure of the EPS to enhance reliability, radically modernize electrical systems through digital technologies, and improve the efficiency of energy production, management, distribution, and consumption.

2. Implementing advanced metering and accounting systems that will improve energy usage by consumers, reduce operational and maintenance costs, and increase awareness and responsiveness to failures in the distribution system.
3. Ensuring an adequate and rapid response from the energy system operator to changes in electricity demand from end-users, thereby enhancing the flexibility of the power system.
4. Implementing strategies for quick fault rectification or system recovery, benefiting from the unique digital platform's ability to quickly and reliably detect, measure, communicate, compute, manage, protect, visualize, and service the full range of elements in transmission and distribution systems. The network is expected not only to alert personnel to the potential negative effects of emergency events but also to maintain a high level of functionality after failures. As a result, the network will localize issues as soon as they arise, preventing them from escalating into major incidents (usually large-scale blackouts). If necessary, the network structure is modified and power flows are redirected to continue supplying electricity to consumers.
5. Integrating renewable energy sources, distributed generation, and modern energy storage systems into the electricity transmission and distribution networks, which is one of the essential conditions for the sustainable development of the energy sector. This stimulates the systemic design, planning, and operation of such power grids.

Thus, the second scenario—renewing the energy development strategy—has become the primary focus today. The implementation of this scenario will also help reduce the negative impact of energy facilities on the environment.

1.2 The Concept of Digitalization and Its Main Objectives

Digitalization is characterized by five key features that distinguish it from automation, which has been widely applied in various economic sectors since the mass adoption of computer technologies.

1. Digitalization places particular emphasis on economic activities, commercial operations, and professional interactions built on new principles using information and communication technologies.
2. One of the main tasks of digital technologies is to radically reduce the rapidly growing costs of market transactions by replacing machines with inefficient operations that require routine human participation. Thanks to these trends, large costs of integration, coordinated work and commercial interaction of millions of new market participants can be radically reduced using digital technologies and platforms.
3. Digitalization involves the creation of a new business model for the interaction of economic entities and the use of digital models that combine the physical and digital worlds. Such a combination is possible when smart machines begin

to form and use digital models of the physical world, which will ensure the independence of decision-making by machines in real time.

4. Digitalization is the emergence of a new type of activity with a large share of scientific, technical and social creativity.
5. Digitalization is a qualitative change in the EPE, consisting primarily in changing the ways of organizing economic relations and leading to the effective involvement of millions of new entities and the smart machines behind them in circulation. Digitalization depends primarily not on technical systems and internal business processes, but on relationships between people, companies and institutions. In this, it is fundamentally different from the automation of production, technological and management processes, which is the greatest achievement of the second half of the twentieth century.

The EPE in developed countries is undergoing a fundamental transformation of its infrastructure to address the problems identified in Sect. 1.1. These changes are aimed at making the electrical grid smarter. For example, the process control systems (PCS) of power plants, dispatch management systems, automation of active energy devices, and metering and documentation systems—all represent important and relevant directions for the evolutionary development of the EPE [5].

Many experts provide the following definition of a smart grid: "It is a technology that enables the transformation of the electric power industry through the introduction of the philosophy, concepts, and technologies that made the use of the Internet possible at power plants and in electrical grids." It is often regarded as a platform for the strategic development of power grids and the optimization of energy and resource distribution.

The creators of the concept proposed implementing the following technological solutions as the foundation for transforming traditional grids into smart ones:

- active network elements, which allow for changes in its topological parameters when necessary;
- autonomous sensors capable of providing real-time data on the current state of the grid;
- information and analytical systems for data collection, processing, and forecasting the grid's condition;
- automated control of the grid through active autonomous high-performance actuators.

Many of these solutions are already used to varying degrees in various systems (for example, in automatic production control systems), but it is the comprehensive approach underlying the smart grid that allows for a completely new level of automation and precision in controlling the grid. Moreover, the transition to smart grids is particularly appealing due to the new economic model upon which the relationship between supplier and consumer is built when delivering services, along with the associated opportunities to increase profits.

The shift to smart grids is not only a revolution in the electrical power engineering but also a catalyst for creating new industries and driving overall economic growth. Digitalization itself has become an industry that enables investment and supports initiatives fostering both technological and intellectual innovation, as well as economic development, including the creation of high-skilled jobs, enhanced efficiency in energy consumption, and optimized use of energy resources. It is not surprising that governments of several countries (USA, South Korea, Japan, China, Russia and others) see the intellectualization of the EPE as a promising opportunity to remain world leaders in the most important economic sector—the EPE. It is important to note that the implementation of the smart grid concept in different countries follows slightly different ideologies and models. The United States, Europe, and China are particularly illustrative examples in this regard.

The smart grid in **the United States** has been primarily developed to promote energy saving through closer interaction between the power system and end consumers. It is estimated that simply installing a smart meter allows consumers to save up to 20% on their electricity bills by optimizing energy usage. Therefore, in the United States, the development of smart grids focuses on the implementation of smart meters and the creation of advanced metering infrastructure. Significant attention is also given to traditional issues of developing distribution and transmission networks, ensuring their ability to self-recover after failures, and improving overall grid efficiency.

In **European countries**, the smart grid concept clearly emphasizes consumer orientation, which primarily implies simplifying the connection of new users, enhancing grid flexibility to meet consumer needs, and promoting cost-effectiveness and environmental sustainability. A simplified diagram of the implementation of the smart grid concept in the European Union countries is presented in Fig. 1.1.

In **Russia**, the following definition of the smart grid concept is preferred [7]: "It is a concept of the innovative transformation of the entire electric power sector, rather than just its individual functional or technological segments. This transformation involves revising several fundamental principles, goals, and objectives for the development of the electric power sector. The resulting scale and nature of the tasks, as well as the anticipated social, economic, scientific, technical, environmental, and other effects, explain the significant attention being paid worldwide to this area of energy sector development".

However, it is incorrect to assume that these approaches differ significantly, let alone contradict each other. Rather, it is a matter of differences in terms used to describe the model, which ultimately aim to achieve the same goal—solving a number of priority tasks:

- creating conditions for the renewal of production assets under tariff growth restrictions;
- creating conditions for integrating new technologies and developing new business models;
- increasing transparency and simplifying supervisory procedures;
- improving legislation and simplifying legal regulations;
- developing measures for government support of competence centers;
- forming programs for training and retraining personnel for the EPE.

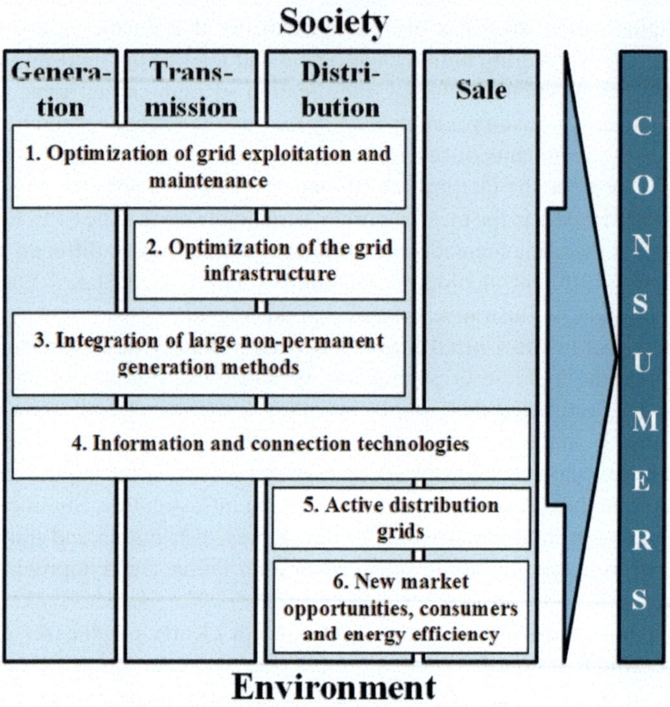

Fig. 1.1 Smart grid development priorities in EU countries

Energy specialists face the challenge of solving digital transformation problems based on a unified information space, which includes:

- systems for monitoring and managing the reliability of power supply;
- platforms for collecting, processing, and using "big data" (including for scientific purposes);
- predictive planning (strategic and investment) and risk planning;
- consumer service.

In essence, the creation of smart grids represents the combination of information technologies, which have already become commonplace in many areas of industrial activity, with powerful electronics and electrical systems. Additional information on the differences between smart grids and traditional ones is provided in Table 1.1.

The implementation of the key provisions of the smart grid concept can be achieved either by improving traditional elements of the energy system or by introducing entirely new ones. However, given the complexity, scale, and comprehensiveness of the required transformation of the energy system, experts have concluded that a realistic approach is a gradual, phased transition during which individual smart grid components will be introduced. While these elements may not immediately deliver

Table 1.1 Technologies used in smart grids for integrated communications

Type of technology	Main components
Wireless technologies	• Multicast communication system • Alert system • Spread-spectrum radio systems • Wi-Fi • WiMAX • Next-generation cellular communication • Time-division multiple access (TDMA) • Code-division multiple access (CDMA) • Terminals for small satellite systems
Other technologies	• Next-generation Internet (Internet-2) • Broadband power line communication (BPL) • Optical fiber-based power supply networks • Fiber-optic cables • Radio frequency identification (RFID)

the full integrated effect expected from next-generation grids, positive results will nevertheless be achieved.

For this reason, in many countries, the implementation of the smart grid concept is undoubtedly a gradual, evolutionary process. For example, in China, only about 3–5% of intelligent (digital) substations were built from scratch, while the rest involved the modernization of existing infrastructure. Gradual modernization should be carried out within a unified, comprehensive project that, for instance, covers a specific territory or segment of the transmission network.

These countries have achieved the widest adoption of smart grids due to a comprehensive approach to organizing the functioning of the energy sector, which includes both regulatory and technological development. Dozens of regulatory documents have been adopted at the governmental level, and institutions have been created aimed at advancing technologies in the fields of energy storage systems, distributed (small-scale) generation, and electricity generation from unconventional renewable energy sources (URES). Significant attention is given to improving mechanisms for attracting investments, supporting and financing projects, and educating the public about smart grids.

1.3 Advantages of Smart Grids Compared to Traditional Grids

A preferred perspective on smart grids is not just what they are, but how they perform better than traditional grids in serving the interests of power companies, consumers, the environment, and the economy [7].

Compared to traditional grids, smart grids have enhanced capabilities to gather and exchange comprehensive real-time information about the power grid through secure and reliable broadband communication channels. This allows operators to have such complete information about the operating condition of the power system that they can make timely decisions appropriate to its state, plan for redundancy, and make decisions to ensure maximum safety, reliability, economic efficiency, and environmental protection. The efficiency of smart grids depends primarily on the identification and study of key performance indicators, the development and testing of appropriate tools, and the creation of proper training programs to equip current and future personnel with the knowledge and skills required for the deployment and operation of this highly complex system [8, 9].

Today's electrical grid was designed to operate as a vertical structure, consisting of generation, transmission, and distribution, and is supported by control systems and devices to ensure reliability, stability, and efficiency. However, system operators now face new challenges, including the need to integrate the existing aging power system with renewable energy sources (RES), distributed (small-scale) generation, and modern energy storage systems, which demand rapid technological changes, as well as an increasing diversity of market participants and end consumers.

Table 1.2 provides a comparison of the characteristics of traditional and smart grids [10].

The tasks addressed in the digital transition for each link of the technological energy chain can be briefly formulated as follows.

Generation (power plants)—Increasing the reliability and efficiency of electricity production using modern intelligent control and management systems (including IT), distributed generation (including from renewable energy sources), and energy storage systems.

Transmission and distribution (grids)—Ensuring the reliability of electricity transmission and distribution, as well as the manageability of the electrical grid through large-scale monitoring and control of operating modes using new tools and technologies (FACTS, Power Management Units (PMU), artificial intelligence, etc.). Transformation (substations)—Ensuring the reliability and controllability of substations by: (a) equipping them with modern electrical equipment, and (b) automating them with the help of advanced diagnostic, monitoring, and management tools based on information and computer technologies.

Consumption—Providing consumers with highly intelligent systems for monitoring and accounting for electricity consumption, regulating energy usage, and managing loads, including during emergencies.

At the same time, the following systems provide technical/technological support for the operation of a smart electrical grid:

- automated accounting and consumer information systems;
- communication and interaction between energy facilities;
- monitoring and control of electrical equipment;
- automation to improve the reliability and fault tolerance of the power supply;
- integration of small-scale power sources and energy storage systems;
- data management;
- management of operational field crews.

Table 1.2 Comparison of existing and smart grids

Preferred characteristics	Existing grid	Smart grid
Active consumer participation	Consumers are uninformed and passive	Informed, engaged consumers; the grid responds to demand and integrates distributed energy resources
Integration of all types of generation and energy storage	Centralized generation dominates; many barriers exist to integrating distributed energy resources	Numerous conveniently located distributed energy resources; focus on renewable energy sources
New products, services, and markets	Limited, poorly integrated wholesale markets; few opportunities for consumers	Efficient, well-integrated wholesale markets; growing new electricity markets for consumers
Power quality for the digital economy	Focus on outages; slow response to power quality issues	Power quality is a priority; various quality/cost options available; rapid issue resolution
Asset optimization and operational efficiency	Insufficient integration of operational data with asset management; fragmented business processes	Greatly expanded data collection on grid parameters; focus on prevention; minimized impact on consumers
Response to system disturbances (self-healing)	Responds to prevent further damage; focus on asset protection after failure	Automatic detection and response to issues; focus on prevention, minimizing consumer impact
Resilience to cyberattacks and natural disasters	Vulnerable to terrorism and natural disasters; slow response	Resilience to cyberattacks and natural disasters; fast recovery capabilities

Combined into a unified platform, these technologies enable a new approach to building power grids, transitioning from the rigid structure of "generation—grids—consumers" to a more flexible one, in which each link, including consumers, can be active. Importantly, when conditions change, the smart grid can automatically reconfigure itself.

The smart grid, first developed around two decades ago in the United States, has already demonstrated several advantages over traditional grids. In particular, it is characterized by an increased level of control over the state of the grid and the integration of power supply functions with strategic planning for high-level energy enterprises.

Figure 1.2 illustrates the sequence of steps/actions that ensure the transformation of a traditional grid into a smart (intelligent) grid [11].

In many countries, in accordance with national legislation, specialized bodies (working groups) have been established to organize and oversee the implementation of activities such as:

- development of standards and protocols,

Fig. 1.2 Actions required for the development of an intelligent electrical system

- ensuring the interconnection between smart grid technologies and the development of infrastructure that enhances the reliability and security of systems,
- ensuring the integration of smart grid technologies with the main processes in the energy system (generation, transmission, transformation, and electrical power consumption), taking into account demand in the global and national electricity markets.

Scientific and educational institutions actively participate in these activities. The involvement of other stakeholders in the intellectualization of the EPE will be briefly discussed in paragraphs 1.4 and 2.1.

1.4 Human Resource Development for the Intellectualization of the Electrical Power Engineering

From the first stage of the implementation of digital technologies, energy companies require employees with new competencies due to the emergence of new types of equipment and new approaches to the design and maintenance of electrical grids and substations. This necessitates the modernization of the existing system for personnel training and retraining. In the context of a shortage of specialists in process automation and IT, energy companies are taking extraordinary measures to address this staffing challenge.

Naturally, the primary role in solving this complex task belongs to higher education institutions (universities). In this regard, they must make significant adjustments to their curricula and the educational process as a whole. At first approximation, the following list of corresponding measures can be proposed:

(1) Development and implementation of breakthrough research programs and innovations in the field of digital transformation of the energy sector and other areas of the economy and social sphere;
(2) Creation of student technology parks and business incubators to accelerate the digital transformation of universities and research organizations, commercialize intellectual property, and transfer technologies to the energy sector;
(3) Updating, developing, and implementing new higher education programs and additional professional programs to accelerate the development of digital technologies in the energy sector, other industries, and the social sphere;
(4) Implementing higher education programs in a networked format, carrying out creative projects with the participation of universities, research institutions, and organizations from the real economy and social sectors;
(5) Improving research efforts in masters, doctoral, and post-doctoral programs;
(6) Improving material and technical conditions for conducting educational, scientific, creative, social, and humanitarian activities at universities, including upgrading their equipment;
(7) Developing the human resource potential of the higher education system and the research sector by ensuring the reproduction of managerial and academic staff, attracting scientists from research institutions and practicing specialists to universities;
(8) Implementing internal and international academic mobility programs for faculty and students, including for joint research projects, creative initiatives, and digitalization projects;
(9) Facilitating the employment of university graduates in energy companies actively engaged in the digital transformation of the energy sector;
(10) Expanding and deepening collaboration between universities and research organizations at both national and global levels, regardless of institutional affiliation;

(11) Replicating the best practices of world-leading universities in training highly qualified specialists for the digital transition of the EPE across other universities;
(12) implementing measures to support young academic staff [5, 12].

1.5 Requirements for University Graduates for Effective Participation in the "Digital Transition" in the Electrical Power Engineering

The textbook authors[1] identified 19 essential knowledge areas and skills that young professionals must master, which can be grouped into four categories: (1) Information Technologies in the EPE, (2) Power System Operation Management, (3) Industrial Networks, Interfaces, and Data Transmission Protocols in the EPE, (4) Information Models of Technical Systems. Below is a detailed description of each category.

1. **Information Technologies in the EPE**:
   1.1. Independent study of principles and methods of setting up information and control systems.
   1.2. Development of screen forms and reports using software tools of information and control systems for automation of business processes.
   1.3. Setting up network infrastructure for interaction of information and control systems.
   1.4. Setting up software of information and control systems in accordance with functional requirements.
2. **Power System Operation Control**:
   2.1. Setting up regulators and limiters in the automatic frequency and power control system.
   2.2. Determining the list of software and hardware packages that are sources of information required for monitoring and control in power systems (PS).
   2.3. Using software and hardware packages to monitor and control power systems.
   2.4. Installing and setting up software and hardware packages to monitor and control EPS operation.
3. **Industrial Networks, Interfaces, and Data Transmission Protocols in the Electrical Power Engineering**:
   3.1. Development of screen forms using software of information management systems for automation of business processes.

[1] Associate Professor S. V. Svechkarev, who has extensive experience in teaching courses on the intellectualization of the electric power industry, contributed to the development of the list of requirements. The authors express their sincere gratitude to him.

3.2. Independent configuration of software of information management systems for solving problems of automation of business processes.
3.3. Configuration of data exchange between information systems using application-level data transfer protocols.
3.4. Software configuration of network infrastructure for organizing data exchange between information systems.
3.5. Carrying out the procedure of testing data transfer, identifying errors and failures in software operation, finding a solution for their elimination.

4. **Information Models of Technical Systems**:

4.1. Analysis of user information needs and development of information models of the subject area based on structural and object-oriented approaches.
4.2. Use of tools for entering and editing information model data (in the electric power industry), checking the correctness of the entered data and correcting errors.
4.3. Changing the structure of the basic profile of the information model (in the electric power industry) and updating it in the information system.
4.4. Setting up software for information and control systems in accordance with functional requirements.
4.5. Exchange of updated information models between control centres in accordance with the rules and regulations.
4.6. Understanding the architecture and technology of operating information systems.

Considering that the concept of smart grids is primarily based on the use of information and digital technologies, the rate of change of which is characterized by the rapid development of technical solutions and approaches, experts consider it appropriate to update the concept at least once every three years (especially in terms of the use of new technologies) [12].

Figure 1.3 illustrates, using the example of the Unified Energy System (UES) of Russia, the changes that have occurred and are projected by 2030 in the balance conditions with the development of smart grids.

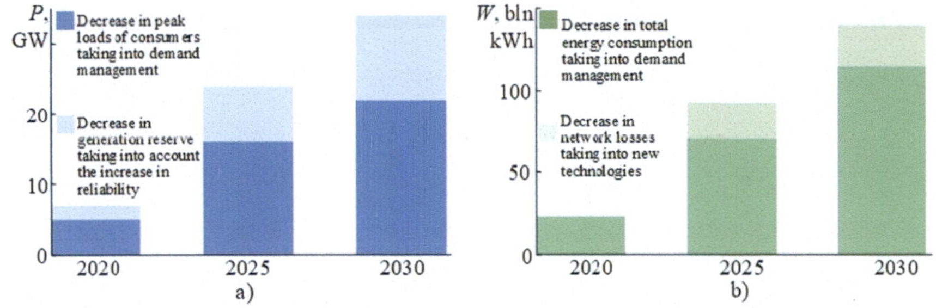

Fig. 1.3 Reduction in required installed capacity (**a**) and energy demand (**b**)

Control Questions and Tasks

1. Name the main problems facing the modern EPE and possible scenarios for its development.
2. What is “digitalization of energy” or “digital transition in energy”?
3. Identify the main differences between a smart grid and a traditional grid.
4. What needs to be done to transform traditional energy systems into intelligent ones?
5. List and explain the essential knowledge and competencies a university graduate must possess to participate in the intellectualization of the energy sector.

References

1. Bauman Z (2007) Leben in der Fluechtingen Moderne. Suhrkamp Verlag, p 287 (in German)
2. Meadows DH, Meadows D, Anders J, Berens I (2004) The limits to growth: the 30-year update. Earthscan, p 363
3. Weizsaecker E, Hargroves K, Smith M, et al (2014) Factor five: transforming the global economy through 80 % improvements in resource productivity. Ernst Ulrich von Weizsäcker: A pioneer on environmental, climate and energy policies. Springer, pp 192–213
4. Lomborg B (2001) The skeptical environmentalist: measuring the real state of the world. Cambridge University Press, p 540
5. Round Table: Smart Grid in Russia: Practice and Perspectives. https://borlas.ru/sites/default/files/pdf/533_document.pdf (in Russian)
6. Economides V, Oligney R (2000) The color of oil: the history, the money and the politics of the world’s biggest business. Round Oak Publishing Company, p 220
7. Kobets BB, Volkova IO (2010) Innovative development of the electric power industry based on the smart grid concept. IAC Energy 208
8. Momoh J (2012) Smart grid fundamentals of design and analysis. Wiley-IEEE Press, Hoboken, New Jersey, p 232
9. The Smart Grid (2010) An introduction. Litos strategic communication, U.S. Department of Energy
10. The Modern Grid Initiative (2008) GridWise architecture council. Pacific Northwest National Laboratory
11. Kinter-Meyer LD, Chassin MC, Kannberg DP et al (2003) GridWiseTM: the benefits of a transformed energy system. Pacific Northwest National Laboratory, U.S. Department of Energy. PNNL-14396
12. Digital Transformation 2030 Concept (2030) Moscow: PJSC "Rosseti". Electronic resource. https://www.rossetivolga.ru/i/files/2019/2/7/kontseptsiya_tsifrovaya_transformatsiya_2030.pdf

Chapter 2
Stakeholder Motives and Participation in the Digital Transition

2.1 Roles and Functions of Stakeholders

To facilitate the understanding and retention of the benefits of smart energy systems, as outlined in Chap. 1, these benefits can be grouped into the following three categories:

1. In the context of restructuring, the deregulated electricity sector allows for market updates based on system constraints as well as seasonal and daily fluctuations in demand. Competitive markets increase electricity flows between regions, which further exacerbates the strain on today's aging energy system and necessitates enhanced real-time control over its operation.
2. Mitigation of network overloads, increased consumer participation in ensuring reliable operation, and reduced uncertainty for investments. This requires improving the network's ability to reliably meet demand.
3. Seamless integration of large power systems with renewable energy producers and distributed (small) generation. The increasing share of energy generated by these sources has a negative impact on the operation of existing power systems, especially if it exceeds 25% of the total generation.

In addition to the energy sector itself (e.g., system operators, managers of different levels, etc.), other stakeholders also contribute to the development of smart grids. Their specific contributions and conceptual understanding of their role in implementing the smart grid concept are discussed below.

In implementing this grid concept, it is essential to identify the stakeholders and understand their potential contribution to the development of smart grids. The benefits of modernizing the power system will be seen across all technology links—from the supply of energy resources to electricity producers to its use by end users.

Figure 2.1 shows the stakeholders and their functions.

In [1], it is emphasized that the economy of the twenty-first century cannot be built on the electrical grid of the twentieth century. Numerous factors that motivate and

V. Y. Ushakov et al., *Digitalization of Electrical Power Engineering*, Power Systems,
https://doi.org/10.1007/978-3-031-95705-5_2

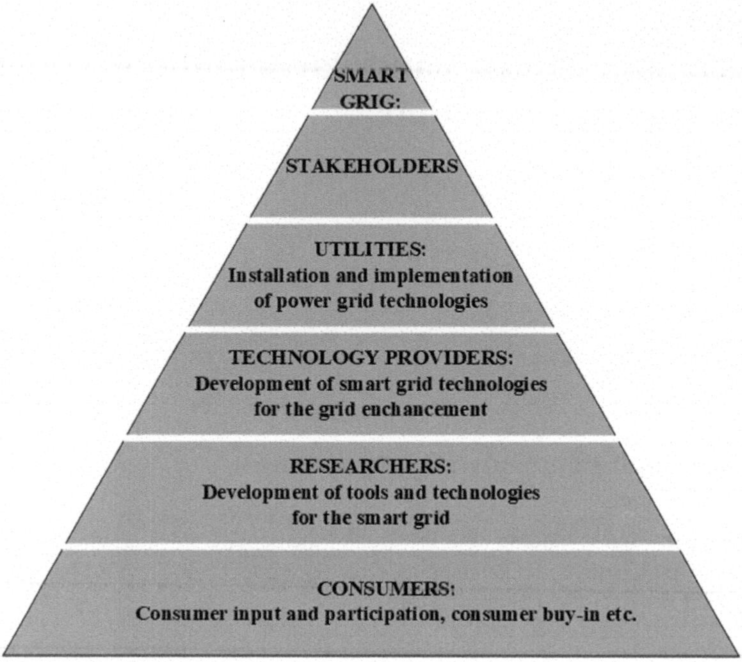

Fig. 2.1 Stakeholders and their functions

incentivize stakeholders to undertake a radical transformation of the energy sector, as well as the challenges addressed in this process, can be categorized into five main groups:

1. **Enhancing the effectiveness of legislative and regulatory frameworks**
   - Introduction of fair market rules for electricity and capacity, benefiting both sellers and buyers.
   - Establishment of reasonable electricity prices and ensuring access to facilitate smart grid capabilities and government regulations that enable capital and operational expenditures for smart grids.
   - A coordinated policy at the federal and regional levels to fully utilize the advantages of smart grids.
2. **Economic Competitiveness**
   - Creation of new enterprises, new business models and "green" jobs.
   - Reducing labor shortages due to demographic problems, rapid growth of all sectors of the economy of developing countries and cross-border labor migration.
3. **Energy Reliability and Security**

- Improving the reliability of power supply by reducing the duration and frequency of outages.
- Reduction of labor costs, such as manual meter readings, on-site maintenance for outages, etc.
- Reduction of indirect costs, such as the use of service vehicles, insurance, damage compensation, etc.
- Decreasing transmission and distribution losses through improved system planning and asset management.
- Securing supplier revenue through more accurate billing and detection and prevention of theft and fraud.
- Providing new revenue streams for suppliers through consumer programs, such as energy management initiatives.
- Reducing capital expenditures by improving network efficiency.
- Meeting national security objectives.
- Enhancing the efficiency of wholesale electricity and capacity markets.

4. **Empowering Consumers**

   - Timely and adequate responses to consumer demand for energy resources and the growing need for uninterrupted power supplies.
   - Enabling consumers to better control their energy consumption with minimal cost and without compromising established living standards.

5. **Environmental Sustainability**

   - Constructive responses to government decisions aimed at increasing the environmental safety of energy facilities.
   - Supporting the integration of renewable energy sources and distributed (small-scale) generation, as well as the development and widespread use of modern energy storage systems.

However, it is important to note that from a funding perspective for large-scale projects like the digital transition, the best option, as shown by global experience, is the public–private partnership model. Expecting energy companies to invest huge amounts of money in the development of smart grids at this stage is unrealistic, since it is still difficult to predict the payback period for such investments. Although this model does not fully meet all requirements in terms of fostering competition within the sector, it remains a viable option. On the one hand, centralized control simplifies the process; on the other hand, various segments of the energy supply chain, particularly generation and distribution companies, operate under market conditions. Thus, there will continue to be interest in both comprehensive and partial integration of these companies. Moreover, this interest will grow as they need to more clearly coordinate their activities to optimize maintenance schedules, equipment utilization, energy inflows, market forecasting, and so on. It is evident that this challenge can only be addressed by establishing a unified infrastructure, including an IT infrastructure. It is anticipated that asset management systems will be the first to integrate, enabling companies to coordinate their maintenance schedules and allowing regulatory bodies to more accurately forecast modernization plans for generation and network infrastructure at the regional level [2].

2.2 Concentration of Efforts by Employees of Scientific Institutions, University Teachers and Engineers of Energy Companies on the Creation of Smart Grids

An example of the effectiveness of concentrating efforts from various institutions to solve the complex problem of energy system digitalization can be seen in the United States, a leader in this field.

2.2.1 Industry Laboratories, Research Institutions, and Universities

A significant portion of the ideas underlying the smart grid concept originated at the Pacific Northwest National Laboratory (PNNL) more than 30 years ago. As early as the mid-1980s, PNNL researchers developed first-generation data collection systems that were installed in more than 1,000 buildings to monitor real-time electricity consumption. A wide range of analytical tools and technologies were developed here, leading to the creation of sensors that are more advanced, diagnostic procedures and equipment designs and operating modes [4].

The Power Systems Engineering Research Center (PSERC), which comprises 13 universities and industrial enterprises, has made a significant contribution to the implementation of the smart grid concept by conducting research aimed at addressing network issues using the most advanced technologies [5]. The primary focus of PSERC's work is the development of new strategies, technologies, analytical capabilities, and computational tools for operations and planning that will support an adaptive, reliable, and stable energy system.

The Electric Power Research Institute (EPRI) and a consortium of universities have developed software architecture as a tool for advancing the technical structure of the energy system through the integration of power supply, communication, and computer control systems. Their software is based on the requirements of integrating data networks and equipment and ensures the functional interoperability of products and systems. This is achieved through methodologies, tools, and standards recommendations for power supply companies to use when planning, specifying, and procuring IT-based systems.

2.2.2 *Technology Companies, Manufacturers, and Suppliers*

One of the world's largest manufacturers and suppliers of hardware, software, IT services, and consulting is the American company International Business Machines (IBM). IBM acts as a system integrator for its smart grid program, which is primarily focused on improving energy consumption efficiency. The smart grid developed by IBM is characterized by enhanced observability with modern data integration and analytics for its control, the integration of the electricity supply chain, and strategic energy consumption planning functions for customers. The network equipment and assets are equipped with or controlled by intelligent IT devices (digital processors). The database and architecture are composed of five main components:

- data sources,
- data transport,
- data integration,
- analytics,
- optimization.

In addition, data dissemination tools are included, such as middleware, publications and subscriptions, portals, and web services [5].

A significant contribution to the creation of IT architecture was made by the aforementioned company Cisco Systems. In recent years, Russian companies such as "Borlas", "EsriCIG", "IBM Russia and CIS", JSC "Klinkmann SPb", LLC "BESK Engineering", and others have played an active role in facilitating the digital transition in the energy sector.

Control Questions and Tasks

1. What features of smart energy systems are most attractive to stakeholders? Who can be considered as such?
2. Name five groups of challenges facing the energy sector that are alleviated by the digital transition.
3. What are the characteristics of the new electricity consumption metering systems proposed by utility companies (electricity suppliers)?
4. What does "public–private partnership" mean in the context of implementing the smart grid concept?

References

1. Mc Donald JD et al (2012) Electric power substations engineering. In: Mc Donald (ed). CPC Press, p 497
2. Round Table: Smart grid in Russia: practice and perspectives. https://borlas.ru/sites/default/files/pdf/533_document.pdf (in Russian)
3. Blueprint for the Future (2009) PEPCO holdings

4. GridWise Architecture Council, Pacific Northwest National Laboratory (2008) The modern grid initiative
5. Taft J (2006) The intelligent power grid. Innov Transform Energy Util Proj 6:74–76

Chapter 3
The Smart Electric Grid

3.1 Key Principles

The power transmission system, which connects all major substations and load centers, forms the backbone of the electric power system (EPS). The goals of transmission companies remain the same: delivering electricity efficiently and reliably at an affordable price. Transmission lines must withstand dynamic load changes and unforeseen events without failure.

The smart transmission system encompasses the network itself, along with self-monitoring and self-healing capabilities, adaptive systems, and predictive tools for generation and demand volumes. The network must be robust enough to handle overloads and disturbances of various natures, responding to them adequately in real time.

Existing transmission and distribution systems were designed before the concept of the "Smart grid" even existed. These systems were oriented toward rapid and inexpensive electrification of countries or regions. However, the requirements for a smart grid are entirely different, making the re-engineering of the current grid inevitable. This engineering effort will focus on improving and expanding the existing grid, carrying out inspection and maintenance measures, preparing for the broader use of distributed generation and large-scale energy storage systems, and developing and deploying an extensive two-way communication system.

The traditional high-voltage transmission and distribution components, consisting of transmission lines, distribution feeders, circuit breakers, disconnectors, transformers, and more, will remain the core of this infrastructure. These components are already technological, and their design can be easily adapted by incorporating smart features through advancements in monitoring, control, and equipment protection.

Strategies to ensure the high efficiency of the smart grid include developing advanced analytical tools and intelligent technologies for network analysis aimed at optimizing power transmission, real-time network condition assessment, and tools

V. Y. Ushakov et al., *Digitalization of Electrical Power Engineering*, Power Systems,
https://doi.org/10.1007/978-3-031-95705-5_3

for modeling the behavior of the electricity and capacity markets. Real-time monitoring of the smart grid is based on the use of modern communication technologies and appropriate devices.

3.2 Key Components of the System Enabling the Smart Grid and Their Functions

According to [2], the system consists of the following key components:

1. Smart metering technologies, which provide faster and more accurate responses to consumer opportunities, such as remote monitoring, time-of-use pricing, and demand management.
2. An integrated, standards-based two-way communication infrastructure, which provides an open architecture for real-time information transfer and control to every endpoint of the network.
3. Advanced management methods, which allow for control of critical components, enabling quick diagnostics and precise responses to any event in “self-healing” mode.
4. Software system architecture with enhanced interfaces, decision support, analytics, and advanced visualization, which enhances human decision-making capabilities, effectively transforming network operators and managers into knowledge workers.

Various organizations involved in smart grid development have proposed different types of grid structures (architectures). One of them, described in [1], is shown in Fig. 3.1.

The system is divided into subsystems based on levels of intelligence and technology, as well as the characteristics of new tools (innovations). It includes electricity generation, transmission, distribution, and consumption.

3.2.1 Energy Distribution Control Systems

Distribution is the final stage of delivering electricity to end consumers. Primary feeders at this voltage level supply small industrial consumers, while secondary distribution feeders supply residential and commercial consumers. At the distribution level, intelligent support schemes provide monitoring capabilities for automation using smart meters, communication channels between consumers and the system managing generating plants, demand management components, and advanced metering infrastructure. The automation system includes modules for self-learning, fault detection, voltage optimization and load switching, automatic billing, feeder restoration and reconfiguration, and real-time pricing.

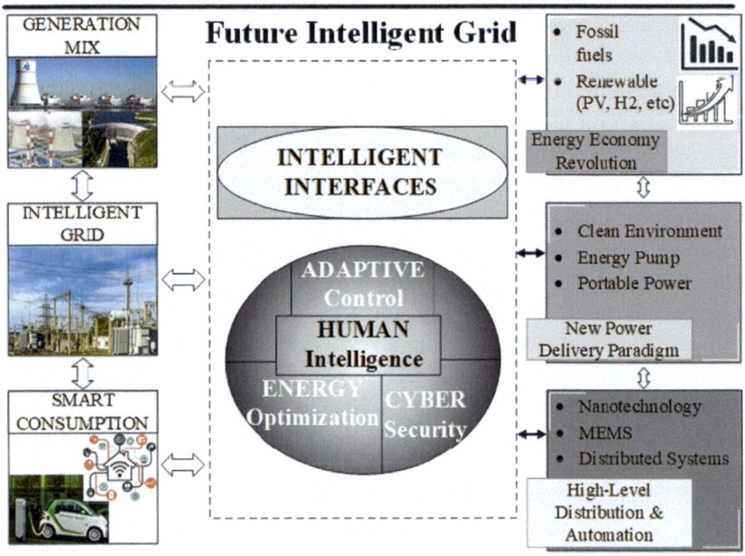

Fig. 3.1 The future of smart grids

3.2.2 *Demand Management Systems*

Demand-side management and energy efficiency systems, designed to respond to changing customer demand, help reduce operating costs associated with expensive generators by postponing the need to expand capacity. Demand management reduces pollutant emissions and waste during electricity and heat production at fuel-burning power plants, lowers costs, and enhances production reliability. These factors collectively influence the load dynamics of the power system. A standard protocol is required to deliver electricity to consumers under two-way communication conditions, along with automated systems and smart meters that transform industrial buildings and homes into "smart" ones. A system to monitor compliance with pollutant emission and waste requirements for energy facilities, as well as energy consumption efficiency indicators, is also necessary.

3.2.3 *New Electrical Power Consumption Metering Systems*

Some electricity generation and supply companies are committing to a radical transformation of their electricity metering systems. Suppliers are moving to advanced metering infrastructure based on open standards (publicly available, non-proprietary technical specifications). These efforts have resulted in continuous improvements in related functions such as customer service, energy efficiency initiatives. Other companies, as well as research and engineering centers, are working on the development of advanced metering infrastructure (AMI) [3]. The proposed functions include investment and implementation of innovative, customer-oriented technologies and initiatives for efficient energy management, enhanced pricing and demand response capabilities, reduction in specific energy consumption and, therefore, a relative reduction in negative environmental impact.

The success of these initiatives depends largely on the quality of the telecommunications infrastructure with its complex architecture and advanced technologies. Although traditional and smart telecommunications networks have the same foundation—high-speed connectivity and compatible communication layers—they differ significantly in their purpose. In traditional networks, communication occurs through human–machine interaction via phones, computers and smartphones. In contrast, the smart grid is essentially a machine-to-machine network consisting of sensors and functional machines that are often not under the direct control of operators and may not necessarily be able to report the status of the network to them. Sensors and measurements in a smart grid can cover the entire process chain: generation, transmission, distribution and end use of electricity. By 2025, the volume of global data is expected to exceed 163 zettabytes (1021 bytes), and by early 2020, about 30 billion devices will already be connected to exchange information.

3.3 Traditional High-Voltage Elements of the Electric Grid

3.3.1 Overhead Transmission and Cable Lines

Smart grid strategies to ensure high operational efficiency at the energy transmission stage include the development of analytical tools and intelligent technologies for network analysis aimed at optimizing power transmission, assessing network status in real time, and tools for modeling the behavior of electricity and capacity markets. Real-time monitoring of the smart grid relies on microcontrollers that manage the energy functions of digital platforms, sensors for assessing network conditions, and communication technologies. Alongside monitoring, testing and diagnostics of the condition of high-voltage equipment in networks and substations remain relevant [3].

3.3.2 Energy Storage Systems

Due to the instability of renewable energy sources and the mismatch between their peak generation and peak electricity demand times, storing energy in volumes sufficient to cover maximum demand remains a pressing issue for the energy sector [4]. Large-scale application of energy storage systems of such capacity would enable the temporal separation of electricity generation and consumption processes, with the following additional capabilities:

- Load management – "discharging" the energy storage device to the load during periods of its maximum values and "charging" during periods of minimum load (at night and on non-working days);
- Power flow management—supplying local loads when the main grid cannot handle it;
- "Hot" reserve—quickly replacing a large generator that has failed in the power system;
- Increased transmission capacity—participating in stability control, voltage regulation, frequency control, and reactive power management, which enhances network stability;
- Load leveling in networks with a significant share of renewable energy sources and distributed energy generation (including those based on fossil fuels);
- Improved power quality—maintaining voltage stability by installing storage systems both on supply feeders and directly at consumers, especially important with highly variable loads;
- Uninterruptible power supply for highly critical consumers;
- Support for consumers with frequent braking and startup modes, particularly in regenerative braking modes (primarily electrified rail transport);
- Reduced requirements for energy generation regulation ranges at base-load power plants;
- Increased efficiency of transmission lines;
- Power reserve functionality during system emergencies.

The digital transition in energy storage systems presents a highly complex challenge due to their wide variety. Energy storage systems can be divided into two broad categories based on the type of energy stored: (1) electrical energy storage systems and (2) potential and kinetic energy storage systems, which are converted into electricity when needed.

The first category includes:

- Capacitive energy storage systems (capacitors),
- Electrochemical energy storage systems (batteries),
- Superconducting inductive energy storage systems.

The second category includes:

- Pumped hydroelectric storage plants,
- Compressed air energy storage systems,
- Gas (air) liquefaction-based storage systems (in "charge mode"),
- Inertial electromechanical energy storage systems (flywheels and super flywheels).

Control Questions and Tasks

1. Name the main components of the two primary parts of a smart grid and provide a brief description of each.
2. Justify the critical role of large-scale energy storage systems in ensuring the efficient operation of electric grids, including smart grids.
3. Name and briefly describe electrical energy storage systems: their operating principles and performance characteristics.
4. Explain the mechanisms and challenges involved in converting potential and kinetic energy stored in respective energy storage systems.

References

1. Amin S, Giacomoni A (2012) Smart grid: safe, secure, self-healing. IEEE Power Energ Mag 10(1):33–40
2. Momoh J (2012) Smart grid fundamentals of design and analysis. Wiley-IEEE Press, p 232
3. Ushakov V, Mytnikov A, Rakhmonov I (2023) High-voltage equipment of power systems. Springer Verlag, p 298
4. Ushakov V (2018) Electrical power engineering: current state problems and perspectives. Springer, p 258

Chapter 4
Testing and Monitoring Systems for the Electrical Grid

4.1 Sensors

Transforming an existing power grid into a smart grid is a large and expensive undertaking, as the grid is one of the most complex technical systems ever created by humans. This transformation requires equipping the existing network with a vast number of smart components (such as sensors, meters, and new communication systems), as well as numerous equipment upgrades.

Although overhead power lines (OPLs) are designed with significant electrical and mechanical strength, their operators often face operational failures due to various adverse effects, including lightning strikes, various types of short circuits, human errors (mismanagement), overloads, falling trees, natural aging (especially of insulating materials), vandalism, and many more. Consequently, OPL faults are one of the most common causes of disruptions in power systems. The time required to restore power supply depends largely on how quickly the fault's location and nature can be identified.

A smart grid requires advancements in diagnostics and monitoring, measurement, communication, and data processing. Furthermore, any new smart grid component must be tested to ensure it meets functionality and performance requirements. Without this, the full potential benefits of a smart grid cannot be realized. Sensors and measurement systems have become integral parts of the network, creating new and complex challenges that must be addressed. Advanced monitoring and measurement technologies enable data collection and transformation into information, which improves power system management and is thus a key factor in implementing the concept of smart grids.

To achieve the "digital transition" in the energy sector, it is necessary to enhance the existing power system infrastructure with additional new types of sensors beyond traditional ones, such as current and voltage transformers.

The list of new sensor types, differing in functions, principles, and designs, includes:

V. Y. Ushakov et al., *Digitalization of Electrical Power Engineering*, Power Systems,
https://doi.org/10.1007/978-3-031-95705-5_4

- Electrical: voltage, current, electric and magnetic fields;
- Environmental: temperature, climate, air composition;
- Mechanical: pressure, stress, spatial displacement.

Sensors enable the detection and timely response to outages, assess the condition of network equipment, eliminate the need for customer site electricity meter inspections, prevent energy theft, allow demand management, and enable optimal power transmission modes.

In the energy sector, measurements based on magnetic and optical effects offer significant advantages, as they allow for non-contact measurement of the required values [4, 5]. This advantage is especially notable in high-voltage devices, as it avoids costly measuring instruments and enhances safety. Current and voltage measurements are essential for monitoring network conditions and organizing protection.

In traditional power systems, current is typically measured using current transformers with magnetic cores. However, these transformers are not suitable for smart grids due to inherent limitations (e.g., non-linearity of the magnetic core, inability to measure direct current, narrow bandwidth, etc.). The smart grid imposes additional requirements on current measurement devices.

Solid-state magnetic sensors have long been used in ammeters, encoders (devices used to evaluate the characteristics of a rotating object), position recorders for linear and rotational motion, tachometers, and more. They are smaller and consume less power compared to measuring coils, fluxgate magnetometers, and more complex devices for weak-field measurements, such as superconducting quantum interference devices (SQUIDs) and spin-resonance magnetometers. Solid-state magnetic sensors operate by converting magnetic fields into voltage or resistance. Measurements can be performed on an extremely small area, further reducing the size and power requirements of the measuring device.

One of the modern types of solid-state magnetic field sensors is the magnetoresistive sensor, which leverages the magnetoresistive effect—the change in material resistivity under the influence of a magnetic field. In recent years, new types of magnetoresistive materials with exceptionally high sensitivity to changes in magnetic fields have been discovered. Theoretical and experimental research, the discovery of novel magnetic materials, advancements in next-generation electronics, and their successful application across various domains have established magnetic sensors as crucial in energy systems, consumer electronics, electrical engineering, and environmental monitoring.

The first generation of such sensors utilizes the anisotropic magnetoresistance (AMR) effect, a quantum mechanical phenomenon whereby the electrical resistance of ferromagnetic conductors changes based on their orientation relative to an external magnetic field.

This effect is widely used in a range of sensors for detecting the Earth's magnetic field (e.g., in electronic compasses), measuring electric current (by measuring the magnetic field created around a current-carrying conductor), traffic monitoring, and more. Sensors based on this effect are produced by depositing a very thin permalloy film. When a magnetic field is applied, magnetic domains "tilt", causing electrical resistance to change by approximately 2–3% at room temperature. In thin films, this effect usually diminishes as film thickness decreases, due to the increased significance of surface scattering, which has limited the application of these sensors and driven the development of new types of magnetoresistive sensors (effectively a new generation of solid-state magnetic sensors) with improved magnetoresistive values. Their development has been accelerated by advancements in spintronics, a new branch of electronics based on the electron's spin degree of freedom. Over the past decade, significant progress has been made in spintronics, particularly through the development and application of thin-film spintronic magnetic field sensors.

Spintronic, which integrates magnetism and electronics, is widely applied in modern electronic devices. In spintronic devices, an applied magnetic field interacts with the electron's spin, determining the electrical current and thus modulating the device's resistance [4]. This facilitates the use of miniaturized, high-speed, and low-cost devices for data recording and transmission, while also advancing the performance of magnetometers. Spintronic sensors are vector magnetic field sensors and can be highly sensitive. For instance, tunnel magnetoresistance (TMR) sensors exploit the quantum nature of electrons. These sensors consist of two magnetic layers separated by an insulating oxide barrier, and the tunnelling current between these layers varies with the angle between the magnetization vectors of the two layers in an externally applied magnetic field. The high magnetoresistance (up to 604% at room temperature) and extremely low magnetic coupling between layers contribute to the very high sensitivity of these sensors. Potentially, the sensitivity of these sensors can reach 1 picoTesla (with some sources reporting sensitivity as low as 1 femtoTesla).

4.2 Testing

Devices used in the power system must undergo the following stages of testing: (1) Initial stage: laboratory tests (typically conducted on individual modules); (2) Intermediate stage: comprehensive assembly testing (including steady-state and dynamic conditions, etc.); (3) Testing for resistance to external impacts, such as pressure, electromagnetic interference, temperature, humidity, mechanical vibration, etc.; (4) Field testing; (5) Final stage: verification of deviations from standard parameters after commissioning (self-diagnostics and measuring instruments).

Testing in a smart grid generally includes all tests conducted in traditional power systems, but these may be subject to different requirements.

In smart grids, testing objects can include:

- Communication and networks,
- Control systems,
- Protection systems,
- Metering systems (smart meters and Advanced Metering Infrastructure—AMI),
- Measurement systems,

- Software.

(The testing of primary power equipment remains the same as in traditional networks and is not covered in this textbook; see [3] for details.)

Compared to traditional electrical grids, smart grids yield a much broader scope of information, including:

- Operability of key power system components (generation, transmission, etc.),
- Dynamics of economic efficiency in their operation,
- Real-time operational parameters of the grid (such as power flow magnitude and direction),
- Geographic and meteorological information, among others.

Testing is critical for smart grids as it ensures that individual equipment components, interconnected in systems, circuits, and substations, are suitable for their intended purpose and meet serviceability requirements. This is typically achieved through testing. Given the trend of using Internet Protocol (IP) technology for real-time control and data exchange, testing is essential to ensure reliability and security. For smart grids to operate reliably, all components must interact effectively.

Smart grid testing may include:

- Testing for compliance with standardized operating modes,
- Power level control,
- Commissioning, activation, and field testing,
- System checks at the acceptance stage,
- Functional testing,
- Electromagnetic interference (EMI) and electromagnetic compatibility (EMC) testing,
- Environmental testing,
- Reliability/availability analysis.

Four primary functions of an advanced sensor system and measurement program in smart grids can be identified: (1) Improving power system measurements and transforming data into information suitable for processing; (2) Equipment condition assessment, network integrity, and support for advanced relay protection; (3) Enabling consumer choice of suppliers and demand response, while also helping to prevent overloads; and (4) Utilizing an advanced metering infrastructure (AMI), which provides an interface between the utility provider and consumers for bidirectional control, real-time electricity pricing, precise load characterization, outage detection, and power restoration.

Existing measurement technologies in smart grids can be categorized by measurement methods and objectives into four types, as shown in Table 4.1.

1. ***Advanced Energy Measurement Technology***: Enhanced energy measurement techniques, including smart meters and other metering systems, contribute to the safety and stability of power supply, high grid quality, fair electricity trading among network users, and various other positive effects. For example, energy

Table 4.1 Measurements performed in the smart grid

№	Measurement type	Application
1	Electrical energy metering	Energy balancing, commercial metering (smart meters, AMI), integration of renewable energy installations, user participation, measurements for the energy market
2	Power quality measurement	Measurement of harmonic contributions in AC and other waveform distortions, DC offset detection
3	Monitoring	State parameter measurement; stability monitoring; monitoring of power devices; monitoring of secondary devices (control, measurement, protection devices, etc.); energy efficiency monitoring; fault location detection; environmental monitoring
4	Measurements for control and optimization	SCADA; WAMS; measurements for relay protection and automation

consumers can achieve energy savings through price signals from smart meters, while the utility provider can prevent electricity theft by tracking local energy measurements.

2. ***Power Quality and Reliability Monitoring*** **via** ***Modern Sensors***: Key indicators include voltage fluctuations, flicker, frequency variations, higher harmonics, voltage surges, sags, and interruptions. Most power electronic devices operate at frequencies of 20 kHz and above, requiring current measurement with a fast response and broadband capability. For instance, current sensors must capture rapidly changing transient signals (within a second or less), which is not possible with traditional current transformers.
3. ***Monitoring Applications***: Monitoring encompasses a wide range of applications, such as measuring system conditions, controlling operational parameters like stability and efficiency, monitoring device conditions (including high-voltage equipment and secondary electronic or communication devices), and environmental parameters (temperature, humidity, etc.).
4. ***Measurements for Network Control and Optimization***: These measurements support protection systems, regulation, dispatching, and planning. Typically, measurements in this category are conducted within a feedback control loop. The Wide Area Measurement System (WAMS), which includes Supervisory Control and Data Acquisition (SCADA) and Phasor Measurement Units (PMUs), is a key component of the smart grid, providing extensive data collection, exchange, and coordinated data management that enhances the grid's efficiency and reliability.

4.3 Monitoring

4.3.1 Monitoring Software Development

The design of monitoring software for fault location involves developing a database, functional modules, a fault location module, and a fault type identification module [6, 7]. An SQL server is used to create the system database, and three primary databases manage data: magnetic field data, transmission line parameters (e.g., voltage, pole locations, wire attachment configurations), and fault information.

All software is developed with Microsoft Visual C ++ 6.0. In the database management module, classes are defined, and properties and methods (such as reading, modifying, and organizing records) are encapsulated, allowing flexible access and database management at any time.

The fault location and type identification module analyses the measured magnetic field signal and determines if an alert signal is needed based on the data. Fault information is then stored in the fault database. Fault location can be identified by changes in the magnetic field's magnitude and direction along the transmission line. First, the short-circuit current level is calculated based on the initialized transmission line data, including voltage, line length, positive and zero sequence parameters. The corresponding magnetic field is then calculated based on pole structure parameters, and the difference in the measured magnetic field magnitude is used to determine the fault location and assess its scale. Finally, the components of the measured magnetic field are analysed to identify the fault type.

The map monitoring module provides users with a visual monitoring interface based on a mapping control (MapInfo MapX), allowing easy integration of mapping capabilities into applications. Maps enable the software to display information in an accessible format, as maps are generally more informative and quicker to interpret than simple charts and graphs. If a short circuit (SC) occurs, the fault location is displayed on the map, and details can also be printed as a report, offering operational personnel a convenient and flexible way to quickly locate faults. In the absence of faults, the current operating state of the transmission line is shown. Additionally, users can review logs and magnetic field waveforms under fault conditions for further investigation, such as identifying hot spots on the transmission line (potential fault points), which is useful for enhancing power system reliability and safety.

The software operates as follows:

- It initializes with predefined transmission line configuration parameters and system parameters from the database. Once initialized, it can monitor the specified transmission line;
- Upon a fault occurrence, the remote terminal system triggers data transmission to the software;
- The location and identification function modules are activated;
- Identified results are displayed on the map background interface in a flashing style.

4.3.2 Non-Contact Fault Location on Transmission Lines Using a Magnetic Sensor

Currently available fault detection and location methods can be divided into two types:

- The traveling wave method,
- The impedance measurement method.

In ***the traveling wave method***, either the transient process initiated by a fault is captured, or probing pulses are introduced into the line, with the reflected traveling wave detected using Time Domain Reflectometry (TDR). Fault location is then determined by time analysis of the traveling wave. As fault signals are heavily obscured by noise, modern signal processing technologies, such as the wavelet method [6], are used to establish the fault location.

In ***the impedance measurement method***, voltage and current are recorded and analysed before and after the fault occurrence, allowing the fault location to be determined through transmission line modelling calculations. Modern relay protection allows data collection via a synchronized phasor measurement unit (a mathematical representation of the signal shape) embedded in a digital relay or fault recorder to pinpoint the fault location.

Unfortunately, these methods have significant drawbacks:

(1) They require expensive equipment (e.g., high-speed data acquisition systems) to be connected to the transmission line;
(2) They yield accurate results only if transmission line parameters are uniform along its length, which often does not match reality. Transmission lines, especially long ones passing through varied terrain with different temperatures and varying ground distances, exhibit different values of resistance, capacitance, and inductance along their sections. In most lines, this heterogeneity results in fault location errors of 1–2% of the line length, but in some cases, it can exceed 5%. For long transmission lines, this translates to an uncertainty of 3–6 km, complicating fault location for repair teams.

Most faults in power systems are short circuits (SCs). The location of an SC can be identified by comparing line current in normal and fault conditions, which can be measured non-compactly by detecting the magnetic field. The strength, direction, and distribution of the magnetic field around a current-carrying conductor contain information on the amplitude, frequency, and phase of the electric current.

For example, a method described in [7] uses a new type of sensitive MR magnetic sensor for transient current measurement. Sensors are placed on transmission line poles; data is sent to a central processing unit where software determines the span in which the fault occurred. Collected magnetic field data can also be used to identify the fault type and even its specific location within the faulted span.

Since MR sensors can independently measure the three components of the magnetic field in three-dimensional space, they can provide information on changes in three-phase current. For radial transmission systems, where there are power sources only at the transmitting end, fault location can be determined by the SC current. Information from the terminal closest to the fault point can be used for qualitative analysis, as it may be affected by the fault. Other terminals can help determine the fault type.

For bidirectional systems with power sources on both sides (e.g., an interconnection line), the fault can be located and analysed by the difference in SC current levels on each side.

4.3.3 Monitoring the Operational State of High-Voltage Transmission Lines with a Magnetoresistive Sensor

Online monitoring, a critical attribute of the smart grid, can track the grid's dynamic state, enabling effective management of transmission lines through a remote monitoring and control system, thereby ensuring the safety and efficiency of the entire power system. Monitoring generally involves tracking the following parameters:

- Electrical: Nominal current and voltage, fault conditions, lightning current, etc.
- Mechanical or Physical: Sag and tension of conductors, pole positions, structural element defects, insulation status, etc.
- Environmental Conditions: Rain, snow, wind, icing, pollution, etc. Accounting for climatic conditions along a transmission line route during its design helps avoid excessive mechanical strength reserves and underestimations of peak electrical loads, contributing to power system efficiency.

The monitoring system itself must meet specific requirements: reliability, safety, ease of installation, stability, and accuracy in detecting critical faults. Non-contact monitoring technology is ideal for meeting these requirements for high-voltage transmission lines, as it avoids installations on live wires, ensuring personnel safety and reducing maintenance workload.

4.3.4 Current Monitoring in the Grid Using a MR Sensors

A new non-contact technology for monitoring current in transmission lines is based on measuring the magnetic field emitted by the lines and solving an inverse problem through current source reconstruction.

Based on the Biot-Savart law, the phase currents and positions of sources determine the distribution of the emitted magnetic field. When magnetic field parameters are measured at several points, solving the inverse problem provides the phase currents as sources of the magnetic field. When the relative positions of the field points and current sources are fixed, source currents can be directly obtained through a deterministic optimization strategy. However, spatial parameters of transmission lines can vary due to conductor sagging or oscillation, making both spatial and electrical parameters of current sources unknown.

To gather data for implementing this inverse problem, several sensors are arranged in a specific geometric pattern, considering that transmission lines are multi-conductor systems and the magnetic field generated around one conductor will overlap with fields generated by currents in neighbouring conductors. Moreover, the Earth's magnetic field and geomagnetic noise are significant, causing the sensor to interpret them as DC and high-order harmonics, respectively. To increase measurement accuracy and minimize interference from other magnetic fields, multiple magnetic sensors are placed around the conductor (or conductors), forming a magnetic sensor array. This new non-contact current measurement method has presented challenges since its inception.

As a result of years of research into signal processing algorithms, noise suppression, and sensor topology optimization, a magnetic sensor array can now be used to measure both DC and AC currents without limitations on conductor shape or the number of parallel conductors.

4.4 Supervisory Control and Data Acquisition Systems (SCADA)

This section of Chapter 4 presents the fundamental principles for designing Supervisory Control and Data Acquisition (SCADA) systems for power utility facilities. It discusses the requirements for such systems based on contemporary technical standards, provides a brief history of SCADA development, and compares various communication systems between objects, highlighting their advantages and disadvantages. The chapter also argues that modern SCADA systems will have standardized architectures and elements to implement the main ideas of the "Smart Grid" concept.

4.4.1 Functional and Communication Requirements for SCADA Systems

A SCADA (Supervisory Control and Data Acquisition) system combines telemetry and data acquisition devices. SCADA facilitates data collection through a remote terminal unit (RTU), transmits it to a control centre, executes any necessary analysis and control actions, and displays information on multiple operator screens or displays. The system allows for any required control actions to be transmitted back to the technological process, including the process of transmitting and converting electrical energy [8].

Initially, data collection systems for process control utilized relay logic. With the advent of more advanced microprocessors and other electronic devices, manufacturers began incorporating digital electronics into relay logic equipment, leading to the development of the Programmable Logic Controller (PLC), which remains one of the most widely used control systems in industry. As the need for monitoring and managing a larger number of devices within a facility grew, the use of PLCs became more prevalent, resulting in more intelligent and compact control systems. While RTUs were initially often specialized devices, PLCs are now frequently used as RTUs.

Advantages of PLC/DCS/SCADA Systems:

- Computers can record and store very large volumes of data.
- Data can be displayed in any format convenient for further use.
- Thousands of sensors distributed across a large area can be connected to the system.
- The operator can integrate real-time data simulation into the system.
- RTUs can collect various types of data.
- Data can be viewed anywhere, not only at the operator's workstation.

Disadvantages of PLC/DCS/SCADA Systems:

- The control system becomes more complex compared to the "sensor-panel" type.
- Operators require diverse skills, such as systems analysis and programming.
- Managing thousands of sensors necessitates dealing with a substantial amount of wiring.
- Operators are limited to information available within the PLC.

In response to the growing demand for compact and intelligent control systems, Intelligent Electronic Devices (IEDs) were developed for PLC and RTU systems. IEDs connect to a computer via a bus, such as Profibus, DeviceNet, or Foundation Fieldbus, and are equipped with the functionality needed for data acquisition and communication with other devices. Each IED can contain multiple sensors, typically combining an analogue input sensor, analogue output, PID controller (a feedback control device), communication system, and memory in one unit.

Advantages of Connecting Computers to the IED Bus System [9]:

- Minimal wiring is required.
- The operator can monitor all necessary information down to the sensor level.
- Data received from the device can include details such as serial number, model number, installation date, and installer information.
- All devices support "plug and play" technology, meaning they are easy to install and replace.
- Device compactness requires less physical space for the data collection system.

Disadvantages of Connecting Computers to the IED Bus System:

- High system complexity requires the involvement of more qualified personnel.
- High sensor costs (partially offset by the lack of need for PLCs).

- IEDs are highly dependent on the communication system.

The design of any system should always be preceded by the development of formal technical specifications and the formulation of relevant technical requirements. These functional requirements reflect the intended system behaviour, which can be expressed as specific tasks or functions that the system must perform. In the case of SCADA, they will include information on points at which system states need to be monitored, the analogue values that need to be recorded, acceptable delays between events and the system's notification of them, the required accuracy of analogue values, and acceptable reliability levels. Functional requirements should also specify the number of remote points to be monitored and managed. During the design process, it is essential to identify all stakeholders in the communication system, beyond the control centre, which may include maintenance engineers and system designers. Communication with the site should be available to them for reasons beyond real-time operational functions.

The analysis of functional requirements should also include formal recognition of the physical, electrical, communication, and security environments in which the communication system is expected to operate. These requirements reflect the following characteristics: possible electromagnetic interference from nearby power systems, the need to identify available communication resources, functional definitions of locations where communication is required, and the identification of security threats to the communication system.

Identifying all elements to be included in functional requirements can sometimes be challenging. In such cases, defining a series of scenarios that detail specific sets of requirements is useful. Once the functional requirements have been formulated, an appropriate communication system architecture can be developed. Communication system requirements should meet the following functional requirements [10]:

- Identification of traffic flows—source/destination/volume.
- Overall system topology.
- Definition of endpoint locations.
- Device/processor capabilities.
- Characteristics of communication sessions/dialogues.
- Addressing schemes.
- Network traffic characteristics.
- Performance requirements.
- Interaction time requirements.
- Reliability/backup/fault tolerance.
- Application service requirements.
- Application data formats.
- Operational requirements (cataloguing, security, and network management).
- Quantitative assessment of electromagnetic interference according to requirements.

Relay Protection Systems perform vital functions to isolate local fault areas within power generation, transmission, and distribution systems to prevent faults from spreading to other parts of the power system. Communication systems are an essential component of relay protection for the entire power grid, providing the necessary data channels for relay operation and control devices. Due to the potential threat of communication loss, relay protection systems should be designed to detect and mitigate communication system failures and avoid repeatedly entering the same fault conditions. The communication system must be designed for rapid, stable, and reliable performance. Factors that can affect relay protection system performance include communication system bandwidth, delays, noise, errors, and overall reliability. Relay systems often require communication delays of only a few milliseconds, frequently disallowing unstable synchronization.

Several data transmission systems can be used for both relay communication and SCADA communication. It is important to note that relay applications often have stricter requirements for speed, latency, and noise than SCADA applications. Power companies use a combination of analogue and digital communication systems, including high-frequency (HF) transmission lines, radio and microwave communication, leased telephone lines, satellite systems, and fibre optics. Each of these systems has characteristics that make them well-suited for specific applications. Their advantages and disadvantages are as follows:

- HF Transmission Lines: Selecting HF transmission lines is typically the most cost-effective option, though it is characterized by limited coverage and low bandwidth. This choice is best suited for inter-station protection and communication with small stations that are difficult or expensive to access.
- Microwave Communication: Economical, reliable, and versatile, microwave communication requires highly skilled maintenance.
- Radio Systems: These provide narrower bandwidth but are useful for communication with hard-to-reach facilities.
- Satellite Systems: Effective for communication with remote facilities, but less suitable where long delays are unacceptable. They are also relatively costly.
- Leased Telephone Lines: Highly effective where reliable communication is needed with a facility served by a standard telephone service. However, over the long term, they tend to be expensive and are generally not the optimal solution where numerous communication channels are required.
- Fiber-Optic Systems: These represent a progressive option. Although installation and maintenance are costly, they are expected to be cost-effective over time. They provide direct communication between facilities and offer very high bandwidth, which is essential for modern data transmission.
- Spread Spectrum Radio Communication: This is an emerging option for communication. Innovations in this field are rapidly advancing, requiring careful comparative evaluation for potential use.

The increasing speed of global networks, combined with the availability of high-precision satellite systems, enables the shared use of communication resources between corporate information, SCADA, and relay protection systems. IEC 61850 provides a communication standard that meets the needs of SCADA systems while also providing functionality for relay protection and automation systems.

4.4.2 Functions of SCADA Systems

Traditional SCADA systems are divided into two categories: "master" and "subordinate" or "remote". Their topology is typically a "star" configuration with a central device. In the past, the master system was a wired device with functional equivalents of indicator lights and buttons, as shown in Fig. 4.1 [11].

Modern systems use a computerized SCADA master device, where remote information is either displayed on an operator's computer terminal or made available to a larger management system through network connections. The substation RTU (Remote Terminal Unit) is either hardwired to digital, analogue, and control points or frequently functions as a subordinate device or "data hub", where connections with intelligent devices within the substation are established using communication channels. Most interfaces in these systems are proprietary designs owned by companies, although protocol-standardized interfaces with RTUs have gained popularity in recent years. In these systems, if configuration or diagnostic information access to the substation is required by other stakeholders, such as engineers or system designers, a separate, often specialized, access route is typically provided through technologies like dial-up telephone lines. With the emergence of communication network protocols typical of the IEC 61850 series, simultaneous communication with users located at multiple remote sites has become feasible. Figure 4.1 illustrates what such a network might look like [11].

This configuration supports multiple users across different sites, allowing them simultaneous access to substation devices for various applications such as SCADA, device administration, system fault analysis, system load measurement, and research.

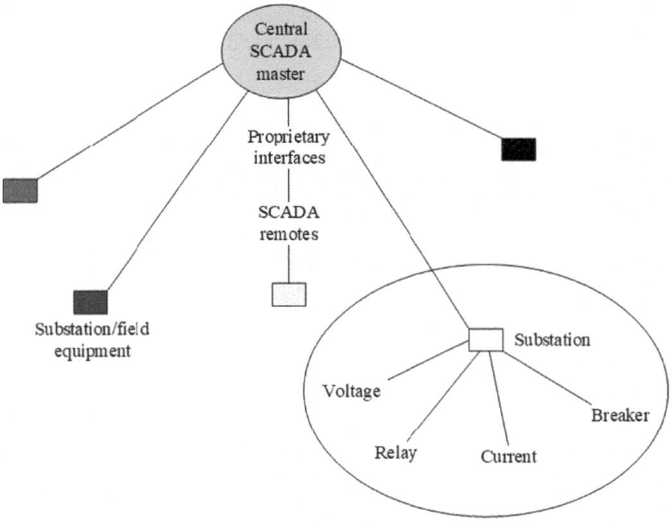

Fig. 4.1 Traditional SCADA system layout

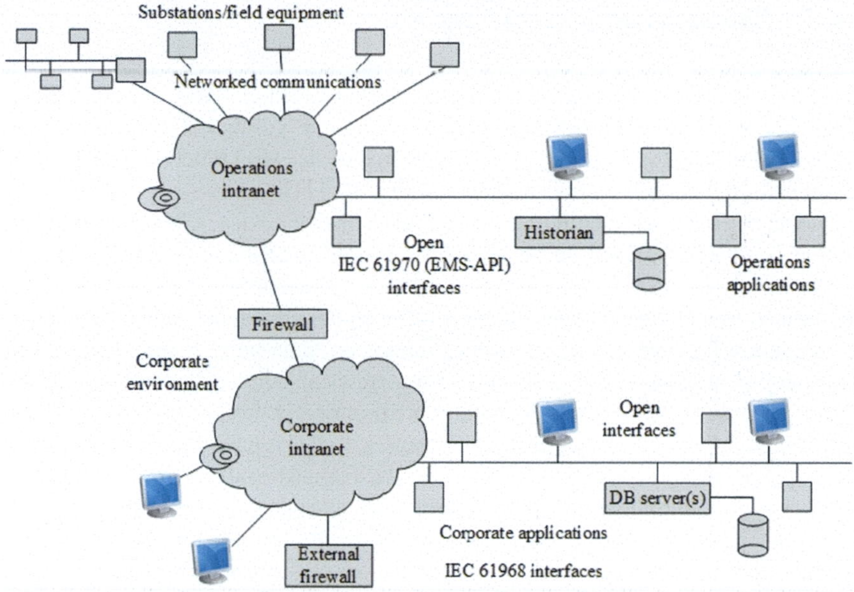

Fig. 4.2 SCADA network communications

Traditionally, SCADA systems transmit information only in real-time, but interfaces based on the IEC 61968 and IEC 61970 standards will allow integration between control centres and enterprise information systems, as shown in Fig. 4.2. Modern SCADA systems can include an archiving function, recording the time when selected status parameters change (outside of a defined deadband) and then storing this information in a database. This archived data can be used to restore the system's state at any chosen moment for performance analysis.

4.4.3 Components of a SCADA System

A SCADA system consists of several RTU (Remote Terminal Unit) devices that collect data and send it to the control centre through a communication system. The SCADA master station displays the received data and allows the operator to perform remote control tasks. Access to accurate and up-to-date data enables optimization of the enterprise's technological processes.

An additional advantage is the ability to ensure more efficient, reliable, and, most importantly, safer operations, resulting in lower operating costs compared to earlier non-automated systems.

In a more complex SCADA system, there are five hierarchical levels [10]:

- Field-level measuring instruments and control devices

- Distribution terminals and RTUs
- Communication system
- Master stations
- IT or data processing computer systems in the enterprise's commercial information technology department

The RTU provides an interface for analogue and digital sensors located at each remote site.

The communication system provides a data channel between the master station and remote sites, which can be wired, fibre-optic, radio, telephone, microwave, or even satellite-based. Special protocols and error detection principles are used to ensure effective and optimal data transmission.

The SCADA master station collects data from various RTUs and typically provides an interface for the operator to display information and control remote sites. In large telemetry systems, subordinate nodes collect information from remote nodes and act as relays back to the main control station.

SCADA software can be divided into two types: proprietary and open-source. Companies often develop proprietary software to communicate with their equipment, which is sold as a turnkey solution. The main drawback of these systems is their excessive dependence on the system provider. For this reason, open-source software is gaining popularity due to its functional compatibility, allowing equipment from different manufacturers to be integrated within the same system.

Citect and WonderWare are examples of open software packages available on the market for SCADA systems. Some packages also include management of all possible enterprise facilities integrated into the SCADA system. Typical components of a SCADA system are shown in the diagram in Fig. 4.3 [11].

The emergence and development of the "Smart Grid" concept have opened up numerous opportunities for enhancing the efficiency of electric network management. SCADA plays a significant role in this as an attractive scheme for remote control and monitoring. At the same time, SCADA technologies continue to evolve, with ongoing efforts to address current challenges in ensuring acceptable levels of reliability, security, and protection in the production, transmission, and distribution of electricity.

Control Questions and Tasks

1. What are the key differences between monitoring and testing?
2. Name the five main stages of testing in the device life cycle.
3. List the primary objects of testing in smart grids.
4. Which parameter forms the basis for most sensors in smart grids?
5. What are the key parameters measured in grid monitoring?
6. Identify the main causes of failures in systems for data collection, processing, and transmission to the control centre.
7. Briefly explain the purpose (functions), structure, and operating principles of the SCADA system.

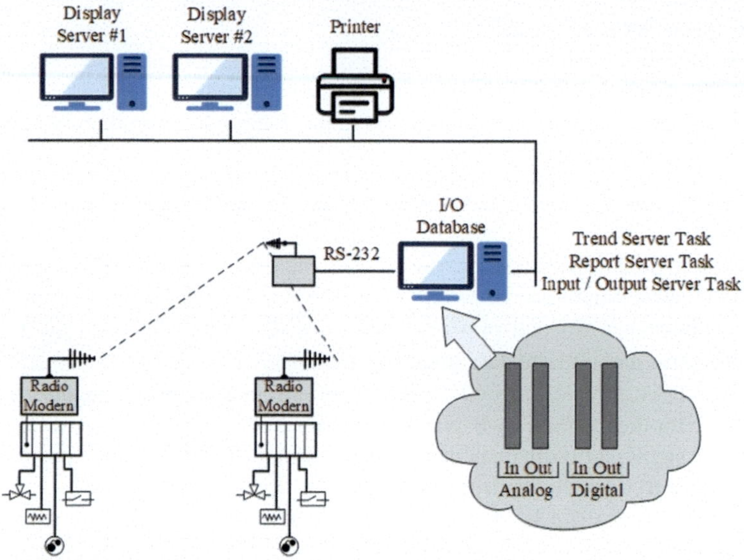

Fig. 4.3 Typical SCADA system

8. List the data transmission systems used at the control centre that are suitable for both relay communication and SCADA communication, along with their advantages and disadvantages.
9. Explain the main principles of SCADA system hierarchy.
10. Describe how SCADA enables the implementation of the "Smart Grid" concept.

References

1. Qi H, Shi J, Jianbo Y, Wei Z (2015) Innovative testingand measurement solutions for smart grid. Wiley & Sons, p 278
2. Fraden J (2008) Handbook of modern sensors: physics, design, and applications. Springer, p 590
3. Ushakov V, Mytnikov A, Rakhmonov I (2023) High-voltage equipment of power systems. Springer Verlag, p 298
4. Freitas PP, Ferreira R, Cardoso S, Cardoso F et al (2007) Magnetoresistive sensors. J Phys Condens Matter 19(16):165221
5. Dill J, Earnshaw R, Kasik D, Vince J et al (2012) Expanding the frontier of visual analytics and visualization. Springer, p 508
6. Abur A, Magnago FH (2000) Use of time delays between modal components in wavelet based fault location. Int J Electr Power Energy Syst 22(6):397–403
7. Huang Q, Zhen W, Pong P (2012) A novel approach for fault location of overhead transmission line with noncontact magnetic-field measurement. IEEE Trans Power Delivery 27(3):1186–1195
8. Gabbar HA et al (2016) Smart energy grid engineering. In: Gabbar HA (ed). Elsevier, p 534

9. Krieg T, Finn J et al (2019) Substations. In: Krieg T, Finn J (eds). Springer International Publishing, p 1079
10. McDonald JD et al (2017) Electric power substations engineering. In: McDonald JD (ed). CRC press, p 524
11. Clarke G, Reynders D, Wright E (2003) Practical modern SCADA protocols: DNP3, 60870.5 and related systems. Newnes, 548

Chapter 5
Digital Substations as a Key Element of Smart Grids

5.1 Main Provisions

Figure 5.1 compares the schemes of traditional and intelligent electrical networks, including substations.

As an important node of the intelligent grid, the DS performs several key functions, such as voltage transformation for subsequent energy transmission and distribution, and management of electricity flows. The creation of a safe and reliable DS is crucial for the development of smart grids. The DS employs advanced, reliable, integrated, and environmentally friendly intelligent devices, and it implements the digitalization of all substation information, a network communication platform, and the standardization of information exchange. The DS automatically carries out key functions such as data collection, measurement, control, protection, monitoring, and more. It also supports advanced functions such as real-time automatic control, intelligent regulation, online analytical decision-making, and interaction. With the advancement of sensor measurement technologies, information and communication technologies, computer technologies, and management technologies, traditional substations are gradually transforming into DS. The latter implement online monitoring of the operational status of primary and secondary equipment, helping to increase the intelligence and reliability of the equipment, improve resource utilization efficiency and production management, and make the operation of substations more cost-effective [1, 2].

This thesis is supported, for example, by the development plan for smart grids in China, which states the following: "Equipment information and maintenance strategies are fully interactive with power dispatching, and full life cycle optimization management is implemented based on real-time condition information. Supporting and central substations must be fully built or upgraded into intelligent substations to achieve unified collection of network operational data, real-time information exchange, real-time control, and intelligent regulation, which ensure the safe and stable operation of power grids and enable various advanced applications". To date,

V. Y. Ushakov et al., *Digitalization of Electrical Power Engineering*, Power Systems,
https://doi.org/10.1007/978-3-031-95705-5_5

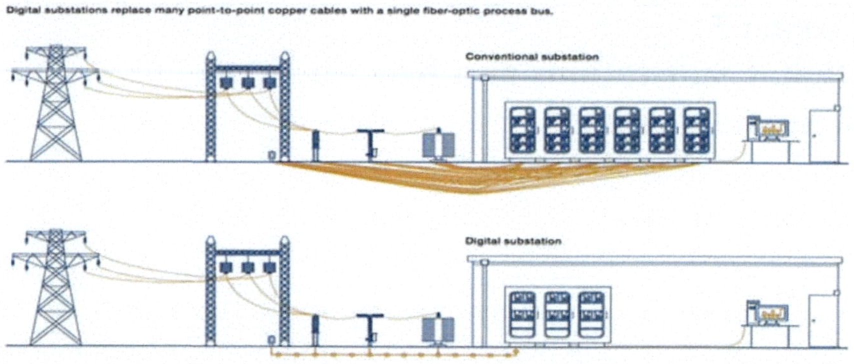

Fig. 5.1 Architecture of traditional and intelligent electrical networks, including substations

a large number of digital substations (DS) with voltage levels of 110–750 kV has been put into commercial operation in China.

The improvement of DS continues. Information in traditional substations is usually isolated within subsystems. With the application of IEC 61850 standards ("Communication Networks and Systems in Substations") and the development of new data collection (sensor) technologies, communication, information, and control technologies, the main priorities for DS are: (1) the shared use of information resources, (2) the integration of various applications and primary and secondary status information into a unified information platform via a single communication protocol, and (3) the implementation of information exchange, automation, and interaction between substations. In the digitalization of grids and substations, key aspects for substations include the creation of intelligent primary equipment, protection and control systems at the substation level, self-diagnostics of equipment, intelligent operational and maintenance systems, and intelligent dispatching systems for electric power systems (EPS).

DS will be safe and reliable in operation, highly integrated into the system, rational in structure and layout, equipped with modern equipment, cost-effective, energy-efficient, and environmentally friendly. Optimization of substation technology and equipment will simultaneously allow for a significant reduction in the physical footprint of the substation and substantially improve safety, reliability, and cost-effectiveness.

To enhance the smart grid's ability to perceive panoramic information from advanced applications and achieve the goal of automation and interaction, DS must take into account technical, economic, and management requirements for the implementation of unified data collection and processing under three conditions: steady state, transient, and dynamic.

In the future, as advanced technologies continue to develop, the improvement of digital substations (DS) will be based on the achievements of the technological revolution and the enhancement of innovative concepts, including new types of equipment, materials, and cutting-edge technologies. A DS equipped with power electronics enables fast and flexible power control, increased transmission capacity, reduced network losses, and decreased short-circuit currents, which can cause network failures. At the same time, data collection and processing at DS are broader, deeper, and more complex than at conventional substations, making information exchange and integration more convenient and faster, and control methods more flexible and reliable.

DS performs important functions related to intelligent primary equipment, full digitalization of information, and standardization of information exchange, and enhanced application interaction, which can be summarized as follows:

1. ***Standardization of Information Exchange***: Unified standardized information models based on the IEC 61850 standard enable information exchange both within and beyond the substation. DS consolidates and simplifies data sources to create a unique and consistent base of information. Thanks to a unified standard and uniform modelling, DS can exchange and share information with other substations, allowing the integration of multiple isolated systems from conventional substations into business applications through shared information.
2. ***Intelligent Primary Equipment***: a) Electronic transformers and intelligent circuit breakers based on optical or electronic principles; b) Digital signal transmission devices and fibre-optic cables (instead of traditional analogue signal transmission systems and control cables). All input and output signals from multi-parameter detection are digital communication signals. The substation communication network is expanded, and on-site sampling data and switch status information can be transmitted across the entire substation or even over a larger area, creating a true digital substation.
3. ***General Digitalization of Information***: Digital substations (DS) can flexibly control both primary and secondary equipment and support bidirectional communication. They can be managed through an information network to meet the requirements of information collection, transmission, processing, and digital output.
4. ***Intelligent Equipment Diagnostics***: In addition to ensuring equipment reliability, DS performs functions such as an autonomous self-diagnostic system for early fault detection and prevention, as well as minimizing power losses caused by equipment failures.
5. ***Interaction with Advanced Applications***: All types of advanced applications, both inside and outside the substation, can interact with each other.
6. ***Commissioning Methods***: Thanks to the Substation Configuration Description (SCD), the design and system integration (Course on Computer Concepts—CCC) will gradually be integrated. SCD contains information about the entire substation model, which is designed and provided by equipment manufacturers for direct uploading. Moreover, with the digitization of device interaction information, substation commissioning is becoming increasingly dependent on software tools that can monitor and verify information. Thus, the automation and reliability of the commissioning process will continue to improve.

5.2 Structure of a Digital Substation

The structure of a digital substation is defined by its functional responsibilities [1, 2]:

- Comprehensive operation,
- Standardized information,
- Coordinated interaction,
- Industrialization of production and commissioning.

A digital substation consists of two primary components- intelligent high-voltage equipment and a unified substation information platform. Intelligent equipment refers to high-voltage equipment that is organically integrated with smart components, featuring digital measurements, network control, state visualization, integrated functions, and interactive information. The main components of intelligent high-voltage equipment include intelligent electronic transformers and smart switching devices. Intelligent components comprise a set of Intelligent Electronic Devices (IEDs) responsible for state measurement and intelligent actuation, including some or all devices for measurement, control, condition monitoring of host equipment (host-equipment refers to any device that provides client–server formatted services as a server over certain interfaces and is uniquely identified on those interfaces), measurement, and protection.

In certain cases, intelligent components are also employed for measurement and protection. A crucial element of a digital substation is the digital relay—a microprocessor-based device remotely controlled by encoded digital alarm signals or immobilizer relay signals (devices providing protection against unauthorized intervention). The digital relay only changes its state upon receiving a unique digital signal from the control unit. Traditional relays are hardwired, with settings manually configured.

The goals of high-voltage equipment intelligence include:

- Enhancing grid reliability by predicting equipment failures;
- Reducing lifecycle costs through intelligent cooling system management (energy efficiency), comprehensive design (land use savings), and online self-diagnosis (reducing asset operation and maintenance costs);
- Optimizing asset utilization.

The first step involves digitizing information about high-voltage equipment. Communication-capable Intelligent Electronic Devices (IEDs) are integrated into high-voltage equipment (theoretically indistinguishable from traditional high-voltage equipment) to enable condition monitoring. Figure 5.2 illustrates the evolution of substations.

The primary motivation behind the development of digital substations is that traditional substations cannot meet the requirements of a smart grid. The challenges associated with automation systems in traditional substations can be outlined as follows:

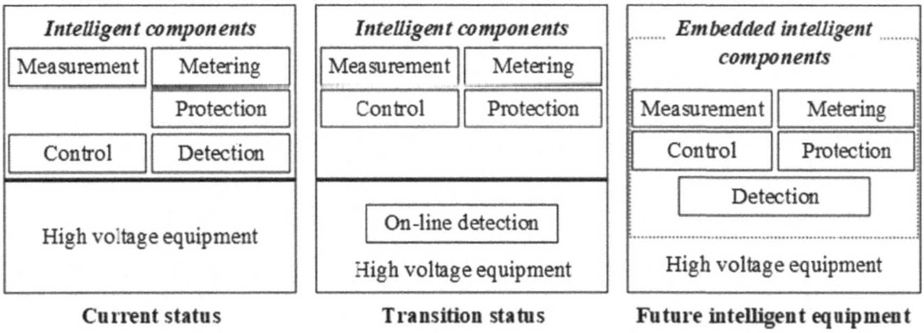

Fig. 5.2 Evolution of substation equipment

- Data inconsistency due to the presence of various data storage and utilization systems within the substation;
- Redundant data collection of the same data;
- High structural complexity and maintenance difficulties due to the diversity of devices;
- Poor compatibility among systems and devices;
- Complex communication protocols;
- Lack of compliance verification and authorization;
- Non-standardized information that is challenging to use.

In traditional substations, there is a strict separation between primary high-voltage equipment and secondary low-voltage devices. In the future, digital substations will eliminate the clear distinction between primary high-voltage equipment and secondary low-voltage devices. All measurement, control, protection, and monitoring devices will be integrated with primary high-voltage equipment. The integration of intelligence into high-voltage equipment will also lead to a rethinking of the functional design of primary high-voltage equipment and secondary low-voltage devices.

Another revolutionary change is in the communication network. Figure 5.3b shows the system architecture of a typical digital substation in comparison with a traditional substation, shown in Fig. 5.3a. In a traditional substation, remote terminal units (RTUs) are rigidly connected by copper wiring to intelligent electronic devices for measurement, control, and protection (see Fig. 5.1). Unlike traditional substations with numerous wired RTUs, manual switching, and limited communication capabilities, digital substations offer a reliable, secure, and open communication network, enhanced data transmission protocols, a flexible human–machine interface (HMI), and compact, user-friendly distributed RTUs. A digital substation can reduce material and labor costs for a utility company while adding additional functionalities and improving substation reliability.

Although wired communication is still widely used, achieving satisfactory control over communication channels is challenging. A digital substation based on the IEC 61850 standard offers significant advantages over earlier designs. GOOSE messaging

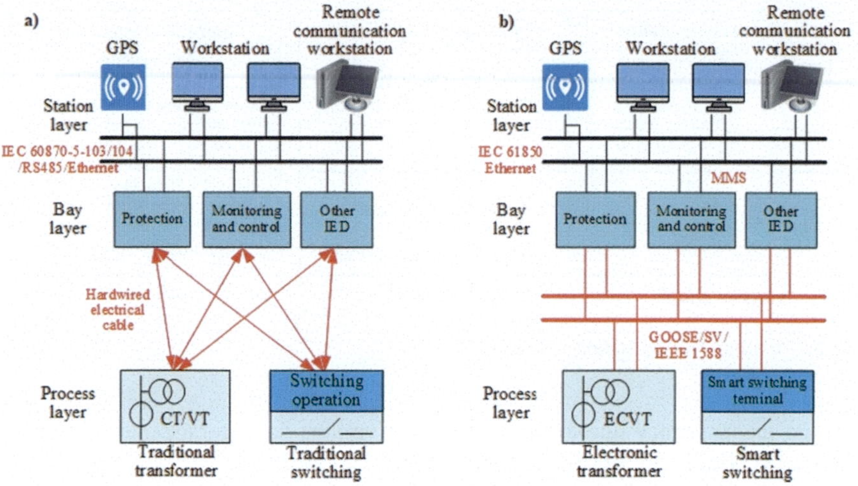

Fig. 5.3 Architecture of traditional (**a**) and Intelligent (Smart) substations (**b**)

also simplifies the wiring within the substation. In practice, only a single cable between the substation's IED devices and the Ethernet switch is required to establish communication between protection and control IEDs, compared to traditional wired connections where a copper wire for each signal would be connected from each IED to all other IEDs in the substation.

IEC 61850, the "Communication Networks and Systems for Substations" standard, defines data stream formats, types of information, rules for describing energy facility components, and protocols for organizing event-based data transmission. This standard was chosen to establish a communication platform for digital substations. Relay manufacturers are expanding their product lines to comply with IEC 61850. The IEC 61850 technology offers a cost-effective solution, reducing the need for wiring between bus sections in switchgear.

A digital substation is divided into three levels: station level, bay level, and process level. The most notable distinctions between a digital substation and a traditional one encompass three aspects: (a) intelligent primary high-voltage equipment, (b) condition-based maintenance, and (c) network-connected secondary devices. The evolution of substation automation systems began with a shift from electromagnetic to digital relays, followed by the adoption of digital communication at the substation level, though proprietary protocols were still in use. In digital substations, the process bus links protection and control devices at the bay level with measuring transformers and switchgear at the process level. This enables the replacement of conventional copper wiring with fibre optic cables, facilitating the transmission of current, voltage, protection signals, and commands via serial networks instead of point-to-point parallel connections (see Fig. 5.1). The process bus also allows the replacement of traditional electromagnetic measuring transformers with new optical current/voltage sensors and the integration of equipment condition monitoring systems.

5.3 Elements of Digital Substations

As noted, digital substations comprise primary (high-voltage) and secondary (low-voltage) equipment. Primary equipment is the high-voltage part of the digital substation, which includes high-voltage insulation, conductors, circuit breakers, and disconnectors, while secondary (low-voltage) equipment is used to control and monitor the primary components.

The manufacturing and operation technologies for primary equipment, such as circuit breakers, disconnectors, and other high-voltage components, are well-established, with primary equipment failure rates lower than those of the associated control equipment. At present, intelligent primary equipment in digital substations is partially in the refinement stage. Some components do not require modifications, while others need only minimal changes (e.g., the installation of sensors).

Intelligent Electronic Devices for Circuit Breaker Management (IED CBM) are used as intelligent interfaces for partial digitalization of primary equipment. Connected to primary equipment via cables, the IED CBM uploads condition information about the primary equipment through GOOSE messaging for real-time management. Meanwhile, the IED CBM, connected to secondary equipment via fiber optics, receives control commands through a downward communication link from the secondary equipment.

The forms of intelligent primary equipment include:

1. With existing actuators (such as spring clamps, hydraulic valves, disconnector drives, grounding switch drives, etc.) and their electromechanical control devices unchanged, the CBM IED and online monitoring unit are installed separately within the control cabinet, as shown in Fig. 5.4.

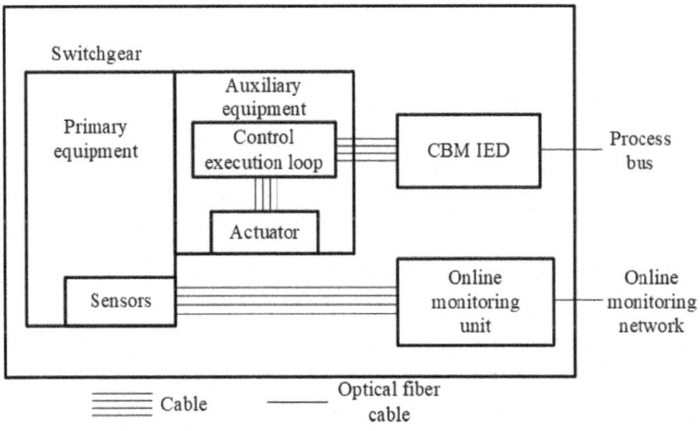

Fig. 5.4 Intelligent electronic devices for circuit breaker management and online monitoring components

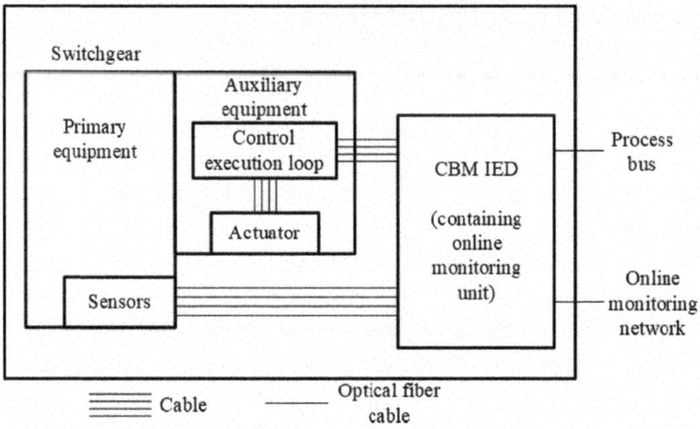

Fig. 5.5 Intelligent electronic devices for circuit breaker management with integrated online monitoring components

2. With the drive and its electromechanical control circuit kept unchanged, the CBM IED is installed within the control cabinet along with an online monitoring function, as shown in Fig. 5.5.
3. Retaining the drive unchanged, the CBM IED performs electromechanical control and directly operates the breaker's trip/close coil, disconnector drive, hydraulic pump, and spring. Additionally, the IED CBM executes an online monitoring function.

The online monitoring technology not only enables the early detection of electrical equipment defects to prevent unexpected failures but also reduces unnecessary power-off testing and allows for certain elements of pre-testing online without power loss. Thus, online monitoring technology enables a shift from regular testing to condition-based maintenance of electrical equipment, saving costs and extending equipment life.

5.3.1 Electronic Current and Voltage Transformers

Current and voltage transformers not only measure current and voltage (and, consequently, electrical energy) but also support relay protection, as well as measurement and control devices [1–3].

Traditional current and voltage transformers are electromagnetic. It is well-known that as power and voltage levels in power systems increase, electromagnetic transformers develop more defects and at a faster rate. Since the 1970s, efforts have been underway to find a new method of measuring high voltage and current that is safe, reliable, and highly efficient. Several types of transformers have garnered particular attention and have been studied extensively. These include Optical Current Transformers (OCT), Optical Voltage Transformers (OVT), Electronic Current Transformers (ECT) utilizing either air-core coils or low-power iron-core coils, and Electronic Voltage Transformers (EVT). Compared to conventional electromagnetic transformers, electronic transformers offer the following advantages:

1. ***Superior Insulation Properties***: In electromagnetic transformers, the magnetic fields of the primary and secondary sides are linked through a steel magnetic core. However, the insulation system is relatively complex, and costs rise quickly with increasing voltage levels.
2. ***Freedom from Magnetic Saturation and Ferroresonance***: Electronic transformers lack magnetic coupling via a steel core, making them free from magnetic saturation and ferroresonance, thus improving stability and reliability.
3. ***Electromagnetic Interference Protection***: For an electromagnetic current transformer, the secondary winding cannot be open-circuited, and for an electromagnetic voltage transformer, it cannot be short-circuited to avoid danger to personnel. In electronic transformers, the two sides are connected by optical fiber, providing electrical isolation. Additionally, since no magnetic coupling exists, the transformer is shielded from electromagnetic interference.
4. ***Wide Dynamic Range and High Measurement Accuracy***: Under normal conditions, the current through the transformer is low. However, during a short circuit, the current rises sharply. An electromagnetic current transformer cannot ensure measurement accuracy over a wide range due to magnetic saturation, nor can it provide precise measurement and protection. Electronic transformers offer a wide dynamic range. The overcurrent can reach tens of thousands of amperes, meeting both measurement and relay protection requirements.
5. ***Broadband Frequency Response***: The converter in an electronic transformer has a wide operating frequency range, with the actual measurement range partly dependent on the electronic circuitry. In contrast, an electromagnetic transformer has a narrow frequency range, with limited response to high-frequency signals.
6. ***Capability for Electrical Energy Measurement, Digitization, Computerization, Automation, and Intelligent Protection***: Microcomputer and digital electronic technologies are widely used within power systems to enhance precision, automation, and intelligence in protective functions.
7. ***Fast and Accurate Signal Transmission* via *Fiber Optic Connections***: Electronic converters, connected by optical fibers, can transmit signals quickly and accurately. The application of power electronics and digital electronic technology meets the requirements for precise measurement and rapid protection action in smart grid environments.

Electronic transformers are divided into two types: active and passive, differing in operating principles, construction, and core technology, as illustrated in Fig. 5.6.

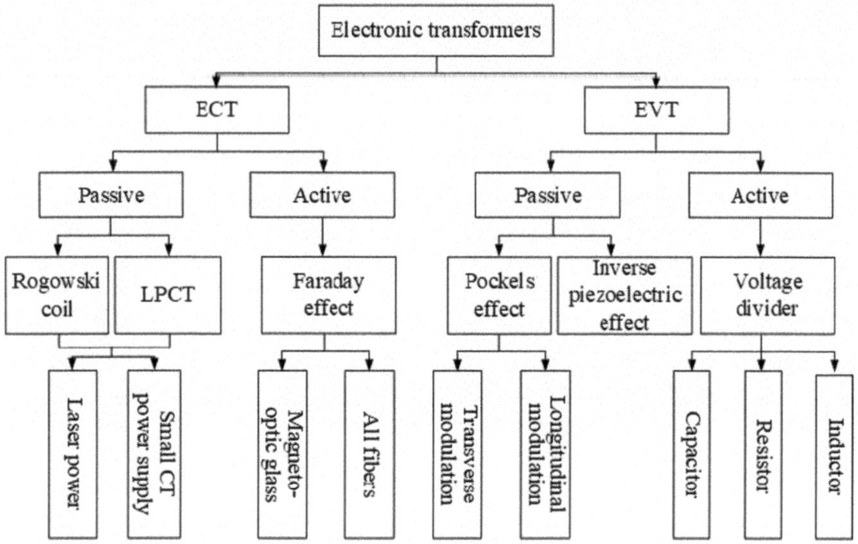

Fig. 5.6 Classification of electronic transformers

5.3.2 Station and Process Buses: Three Levels of Communication

Figure 5.7 shows the three typical levels of digital substation control architecture [1, 2]. At the station level, general equipment such as human–machine interfaces (HMIs) and communication interfaces is utilized. The bay level is more specific to each circuit, employing equipment for each circuit (e.g., protective relays and local control units). Primary equipment (such as measuring transformers, disconnectors, and circuit breakers) is located at the process level. While traditional connections between levels used copper wiring, the adoption of substation communication buses has reduced the need for cables. IEC 61850 enables different manufacturers to produce devices compatible with a single communication bus, allowing for shared information exchange. Status information (signals) about equipment can be distributed across devices from different manufacturers, eliminating the need for multiple connections for status and other information transmission—once a signal is configured for one device, it can be transmitted to others.

As a further enhancement, IEC 61850 specifies replacing copper wire connections between the process level and bay level with fiber optic cables, enabling a transition from analog signals to digital versions. Signals from conventional current and voltage transformers can be digitized using merging units at the switchyard or modern unconventional current transformers utilizing fiber optics and the Faraday effect or voltage dividers to replace traditional voltage transformers. Each of these signals undergoes analog-to-digital conversion.

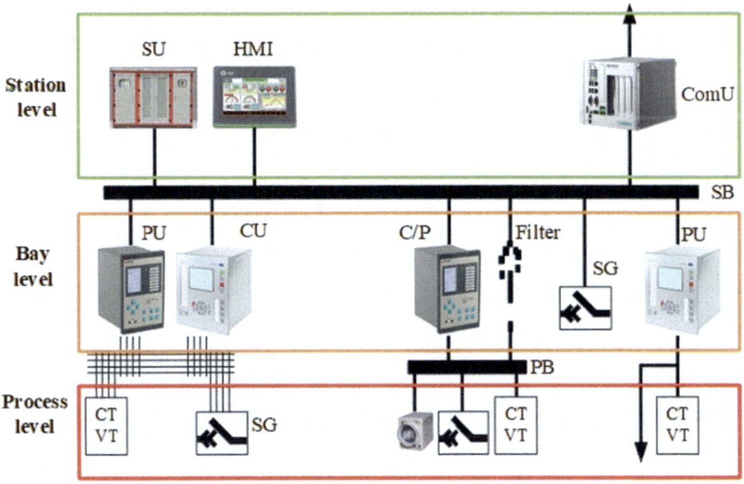

Fig. 5.7 Three levels of substation communication

Trip signals can also be transmitted over the network through GOOSE messaging instead of conventional wiring.

Figure 5.8 presents a block diagram of the substation with various communication levels and possible interconnections. As substation automation equipment has become digital, so too has the commissioning and testing equipment. Modern testing equipment can be programmed to perform specific test sequences, reducing time and costs. This is further supported by the use of the process bus and "sampled values" of primary current and voltage.

Many networks still operate with electromechanical relays that have been in service for 40–50 years after installation. In contrast, digital devices typically have a lifespan of around 10–15 years due to component obsolescence and challenges with ongoing support from manufacturers. This means that, over a typical 40-year substation life, the substation automation system will likely be replaced at least once, if not twice, over the lifespan of the primary equipment. This can be at least partly offset by the fact that the relative cost of digital equipment and development is generally lower than that of electromechanical devices. To avoid future issues, it is advisable to design substation secondary equipment with easy replacement in mind at the construction stage.

Adherence to IEC 61850 guidelines facilitates equipment modernization. The reduction in copper connections and the use of Ethernet connections mean that new devices can be connected to the substation buses, configured, and tested within the substation environment before replacing older units. In a digital substation, a far greater amount of equipment across various functions can be monitored, collecting additional data about substation conditions. Trend analysis on equipment condition allows for maintenance based on actual condition rather than a fixed schedule.

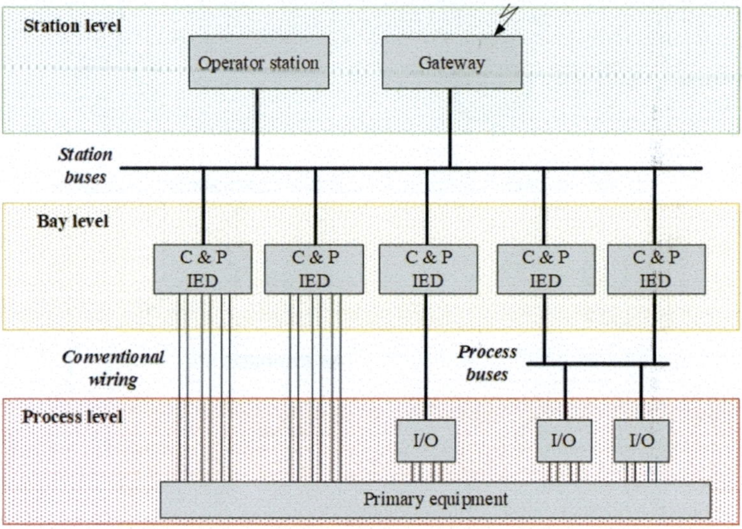

Fig. 5.8 Block diagram of substation with communication levels

New methods are also suitable for assessing the dynamic stability of power lines by measuring wind speed.

5.4 Integrated Information Platform and Advanced Applications

As shown in Fig. 5.9, the integrated information platform provides standardized information access interfaces for intelligent applications and remote systems by gathering all SCADA data, protection information, recording data, measurement data, and online monitoring data from the substation. This platform effectively addresses issues such as an excess of systems and interfaces, poor data exchange, and synthetic application of substation data. It also meets digital substation requirements for data digitization, function integration, structural compactness, and status visualization.

Based on the information integration platform, advanced functions can be performed, such as one-touch sequence management, source-level maintenance, intelligent alarm and fault analysis, intelligent billing, and load optimization management. These functions are briefly explained below.

1. *One-Touch Sequence Management*: The platform's primary function in sequence management is to automate operational tasks according to the standard operational order of the substation. The platform can execute several stages of management simultaneously. Sequence management automatically checks the

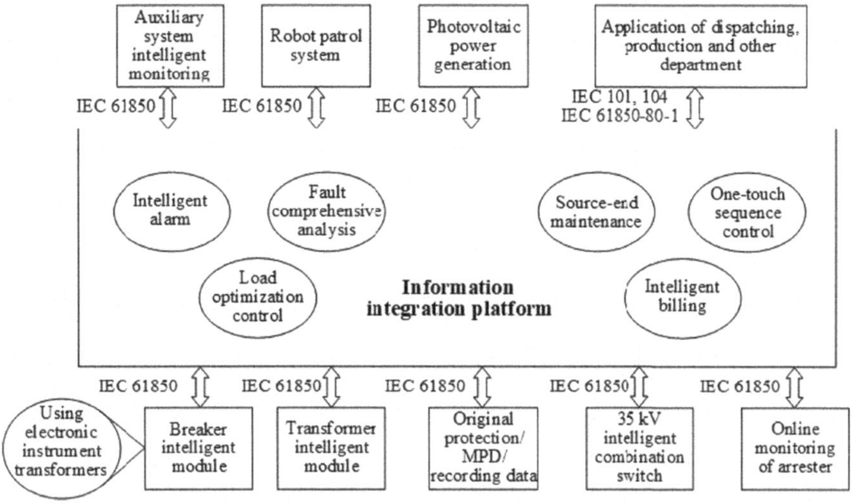

Fig. 5.9 Integrated Information Platform for Digital Substations

lockout logic to prevent improper operation before each step, with functions for interruption and emergency stop.

The sequence management system combines one-touch control with a video system through an intelligent patrol robot. When interacting with primary equipment, the video system directs the camera to the device, and the device status is assessed through image recognition technology, which then relays information to the sequence management system. The sequence system assesses equipment status comprehensively based on measurement information and data from the video system to ensure reliable operation.

2. *Source-Level Maintenance*: As a data collection terminal of the centralized control system, the substation must provide various self-descriptive configuration parameters that can be set through a single tool directly at the substation during maintenance. Furthermore, the substation can generate standard configuration files, including parameters such as the main connection diagram, network topology, and other data models.

   The primary electrical diagram and graphic files of the substation automation system must be available in the dispatch and central control system in a scalable vector graphics (SVG) format, as shown in Fig. 5.10. The display relationships between the substation model (IEC 61850) and the main dispatching model (IEC 61970) are established to create primary maintenance data models and graphical images on the substation side, used in real-time by various automation systems in the dispatch center to reduce maintenance workloads and ensure consistency across substation models and data. Source-level maintenance significantly reduces workload in the dispatch system and eliminates the need for verifying substation operation and dispatch data accuracy.

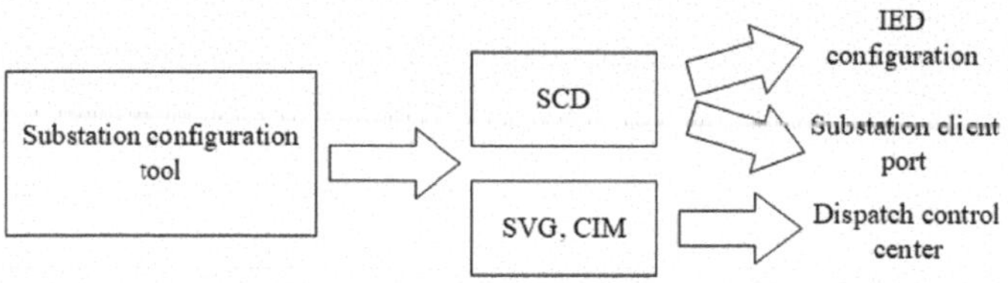

Fig. 5.10 Source-level maintenance

3. *Intelligent Alarm and Fault Analysis*: The intelligent alarm system pre-processes alarm/threat information and creates an expert system database for fault processing. This system synthesizes alarm data, brief fault reports, signal records, and other information and integrates the intelligent alarm system with the human–machine interface.

   Fault analysis software based on the information integration platform provides fault diagnosis, fault localization, equipment monitoring and assessment, harmonic analysis, signal processing, and other fault analysis functions, offering a comprehensive fault information analysis.
4. *"Intelligent" Substation Access System*: The intelligent access system can automatically issue various types of operational entry passes based on the official and actual operating modes of the facility, allowing operators to generate work permits with a single click.

   Key to the intelligent access system is the development of sound operating rules, making access smarter and simplifying maintenance. When setting these rules, operational procedure and substation operation restrictions are considered. A real-time interface reads device attributes and status when generating operation permits. This information is then matched with the selected task and its execution rules to generate a suitable action sequence. If intelligent access is selected, the system automatically generates the necessary operation list based on the device's actual operating status and rule database.
5. *Intelligent Load Management*: The Voltage Quality Control (VQC) module is embedded within the substation controller (remote terminal unit) system software. Real-time data and equipment parameters required for its algorithm are fully based on the information integration platform. VQC is controlled by the main station system, and its operating voltage and target reactive power settings are obtained from the dispatch center. The optimized control equipment participating in VQC can automatically decide whether to engage in VQC regulation depending on equipment verification status. Based on real-time data and the current optimization target, VQC automatically selects the optimal control method for the relevant equipment, adapting to different operating modes. All actions are performed within the substation, and the results are sent to the dispatch control system.

   Overload shedding is also embedded in the substation controller system software as a program module controlled by the dispatch system. Real-time data and algorithm parameters are incorporated into the integrated information system, with priority actions also set by the dispatch system. In case of system failure and overload of main substation transformers, the advanced application will automatically disconnect lines with the least critical loads.

5.5 Intelligent Condition Monitoring of Primary Equipment

Modern trends in the digitalization of electric power are shaping a next-generation power system characterized by a continuous increase in data volume, which allows parts of the power system to be viewed as conditionally observable. However, big data is naturally generated only for newly implemented elements of the power system, while other elements remain insufficiently observable due to the lack of monitoring systems and tools for assessing the complexity of interconnected processes in power generation, transmission, and distribution, as well as the economic inefficiency of installing measuring devices on every piece of equipment.

If we consider high-voltage equipment at power plants and substations as a system and recognize that each unit of equipment comprises elements connected by physical processes, the current technical condition of the equipment can be viewed as a system of diagnostic parameters and their values at a specific point in time. Each unit of equipment is a complex technical system composed of multiple elements with interconnected processes of varying nature, complicating the assessment of the current technical condition of high-voltage equipment. Managing a complex technical system requires determining its current technical condition, including all subsystems and elements, predicting their changes, and managing the technical condition of individual units of power plant and substation equipment, as well as major components and the condition of the entire facility.

The challenge of improving the accuracy of determining the current technical state of equipment and predicting its changes is closely tied to the lack of retrospective and current diagnostic data. Consequently, most power companies adopt a strategy of scheduled preventive maintenance and repairs with periodic inspections set in regulatory documentation, regardless of equipment condition. This strategy can lead to unnecessary maintenance labor or, conversely, to delayed detection of defects or malfunctions.

Thus, in the global power industry during the 1990s, a scientific and technical challenge emerged: the transition to condition-based maintenance and repair. Numerous scientific and industrial studies have addressed this issue. Condition-based maintenance and repair are based on three main stages: analyzing retrospective diagnostic parameters, determining current values, and forecasting future changes. At each stage, diagnostic information must be reliable, objective, complete, and up-to-date.

Solving this issue requires new approaches to big data collection and analysis technologies, which would not only improve the accuracy of determining equipment condition but also extract actionable insights and accurately generalize findings, identifying hidden patterns in diagnostic parameters and causative factors. This approach could reveal correlations at the equipment element level, as well as interactions between the technical conditions of individual equipment units.

Intelligent hardware and software systems, as integrated networks of physical (robotic systems) and computational (software) components, offer a promising tool for collecting and analysing big data in equipment condition management. The use of machine learning algorithms in these systems enables resilience to outliers, the ability to analyse fragmented data, work with various data types, and provide automatic system training. Therefore, these systems are effective automated tools for enhancing the accuracy of high-voltage equipment condition assessment and reducing incidents in power facilities.

Establishing a scientific and methodological foundation for automated diagnostics of high-voltage equipment at open distribution facilities in power plants and substations is a relevant research area, facilitating the shift to condition-based maintenance [4, 5].

Wear and defects in power equipment show early signs manifested in gradual changes in electrical, physical, chemical, and other characteristic parameters [6]. Thanks to sensor, computing, and communication network technologies, characteristic equipment parameters can be promptly obtained, analyzed, and processed by an expert system capable of assessing equipment reliability and estimating remaining service life. This enables the early detection of potential failures, enhancing power supply reliability. Online monitoring can continuously track and evaluate operational power equipment, providing the necessary basis for assessing equipment condition.

Transformers, circuit breakers, and other primary equipment undergo regular maintenance and commissioning. Preventive tests (offline) are conducted after scheduled power shutdowns to gather information and assess whether primary equipment is fit for continued operation. This preventive method requires power shutdowns, and improvements are needed in authenticity and real-time response.

As technology advanced, an online monitoring system for certain parameters was developed. For example, several transformer parameters, such as dielectric losses, core current, dissolved gas, partial discharge, micro-moisture content in oil, hot spot temperature, and winding deformation, can now be monitored online. Today's online monitoring technology has addressed some issues associated with offline testing, and experience in its application has grown, significantly enhancing operational efficiency. However, numerous challenges remain, such as incomplete testing parameters, poor compatibility with equipment, and technical implementation complexities. Online monitoring technology still cannot fully reflect real-time equipment operation and lacks standardized protocols. Consequently, it does not fully meet the requirements for building smart grids to enable online monitoring of substations.

5.5.1 Point Measurement of Transient Magnetic Fields in Substations Using a Magnetoresistive Sensor

The development of smart grids can be hindered by electromagnetic interference (EMI) in substations, as modern power systems increasingly place secondary systems (such as intelligent electronic devices like solid-state protective relays and microprocessor-based control units) not in control rooms but in switchyards, adjacent to high-voltage equipment [7]. This makes it essential to ensure the resilience of transient electromagnetic field measurement systems to interference caused by switching operations or lightning.

In transient modes, secondary systems installed at substations are subjected to powerful electromagnetic waves composed of magnetic (TMF) and electric (TEF) components. Measuring the magnetic component (TMF) is more challenging:

Firstly, magnetic fields are harder to shield compared to electric fields. Many cabinets in traditional substations, previously located in control rooms, are now situated in switchyards where they lack adequate TMF protection.

Secondly, Current surges (TMF) pose a greater threat than overvoltage (TEF), as substations are typically well-protected against overvoltage.

TMF measurement is complicated by their broad frequency range [8]. Available TMF recorders can generally only measure interference levels without analysing the high-frequency component, which is crucial for diagnosing actual faults in the power system. Finding a solution that allows for TMF recording and analysis over a wide frequency range is critical [9].

In addition to measuring TMF levels, it is necessary to assess their impact on communication and control devices (secondary circuits) used at substations. A measurement system for TMF must have the following capabilities: (1) high spatial resolution TMF measurement; (2) a sufficiently wide bandwidth and dynamic range; (3) the ability to correlate TMF with power system events.

A traditional solution for TMF assessment in conventional power systems is a search coil, which performs well at high frequencies. However, in smart grids where TMF needs monitoring at low frequencies as well, the large size of these coils makes point measurements difficult. A system meeting these requirements has been developed based on MR sensors, as described in [8].

Point measurement systems for TMF with a measurement range of up to 6 Gs (with a resolution of 0.02 Gs) and a bandwidth from DC to 5 MHz have been developed, covering most transient processes in substations.

5.5.2 Online Monitoring of Power Transformers, Gas-Insulated Switchgears and Surge Arresters

The power transformer is one of the most important and expensive devices in the power system. Its safe operation is crucial for ensuring the reliability of the power supply. To improve operational reliability and reduce economic losses from failures and accidents, regular preventive testing of transformer insulation is necessary [6]. However, if preventive testing requires power shutdowns, it disrupts normal power supply. Therefore, online monitoring of transformer operation has become a priority. The development and widespread application of online monitoring technology is foundational to condition-based maintenance, which will play an important role in power systems.

Currently, transformer monitoring primarily includes the following aspects:

- Online monitoring of partial discharges in transformer insulation.
- Online monitoring of on-load tap changer (OLTC) operation.
- Online monitoring of transformer bushing insulation.
- Online monitoring of transformer oil and winding temperature.
- Online monitoring of water content in transformer oil.
- Gas chromatography monitoring of transformer oil.

Monitoring, maintenance, and overhaul of switching equipment are essential for the safe operation of power equipment and systems. During condition monitoring, physical and chemical values reflecting equipment operating status are determined using various sensors and measuring instruments. Online or autonomous monitoring, measurements, testing, and evaluation of operating characteristics of primary equipment and related components help determine if equipment is in normal condition.

In gas-insulated switchgear (GIS), there are no exposed live parts except for bushing insulators. SF6 insulation provides high reliability and requires less maintenance. Additionally, external diagnostics and inspection (without dismantling) can indicate whether dismantling and maintenance are necessary.

1. Partial Discharges: High insulation quality in GIS is essential for reliable and safe operation. Metal particles, dust, and moisture within GIS equipment increase the likelihood of failure. Conductive particles in GIS insulation can cause partial discharges, accompanied by specific phenomena such as abnormal sound, vibration, glow, and gas decomposition. Therefore, partial discharges are a key aspect of GIS condition monitoring.
2. SF6 Monitoring: The properties of SF6, used for insulation and arc quenching, are critical for GIS. Therefore, it is necessary to monitor gas pressure, seal integrity, and micro-moisture content.
3. Monitoring Mechanical Characteristics of Circuit Breakers: For online monitoring of current waveform in opening and closing coils, a compensated Holzer current transformer is used. The current waveform is compared to a standard waveform to detect anomalies in the circuit breaker's mechanical characteristics.

Mechanical characteristics (stroke, opening/closing time, average speed, etc.) of the circuit breaker are determined using a non-contact optical barcode reader, followed by diagnostics. Monitoring SF6 pressure and checking the operation of the compressor station are also important.

In online surge arrester monitoring, the current in the arrester ground lead (sometimes referred to as leakage current) is used to assess its condition. Changes in current magnitude and waveform are converted to changes in the optical pulse frequency for subsequent decoding. Key issues, such as measuring surge arrester leakage current, passive sampling during transmission, high-voltage insulation, remote data transmission, and over-limit leakage current alerts, have been addressed. Automation of surge arrester insulation status monitoring has also been implemented at unmanned substations.

5.6 Digital Substation Protection

Next-generation digital substations use secondary equipment to address challenges such as environmental protection, EMI shielding, and enhanced reliability of locally installed equipment. To reduce device numbers and simplify secondary wiring, multi-functional CBM IED (Circuit Breaker Management Intelligent Electronic Devices) monitoring and control units are employed. The integration of digital measurements is encouraged to further improve data exchange and simplify the measurement system.

The advanced substation protection system utilizes a hierarchical approach, synthesizing panoramic information on the power grid and creating hierarchical protection management systems consisting of temporal, spatial, and functional measurements. This approach aids in establishing a more robust power grid security framework. The hierarchical protection system for the next-generation digital substation is shown in Fig. 5.11.

For local protection of a single asset, independent information about that asset is used. The component with a detected fault is reliably and quickly isolated. The station protection management system gathers information on protection operation and alarms and directly sends instructions, bypassing locally installed protection.

In the temporal dimension, local protection for various primary protection types operates with no time delay (20–30 ms). To ensure selectivity and reliability, protection speed is sacrificed (0.8–1.2 s). Station-level and global protection can utilize comprehensive information to accelerate local backup protection (0.3–0.5 s). Protection and stability control at each level interact with each other, improving relay protection efficiency and stability control. Their relationship is shown in Fig. 5.12.

From a functional perspective, to enable rapid and reliable disconnection of a faulty component in local-level protection, information from individual components can be used independently. To optimize protection management configuration and enhance protection and control performance, centralized protection of the asset utilizes all station information, providing flexible, adaptive bus protection, failure protection, component backup protection, and load-shedding control functions.

Local and station-level protection are independent of external communication channels, allowing protection functions to operate even if communication channels

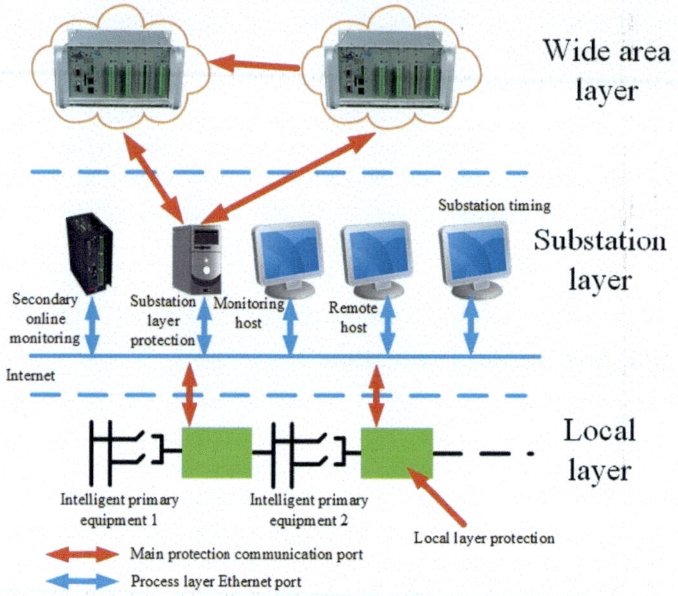

Fig. 5.11 Schematic diagram of the hierarchical protection system in a digital substation

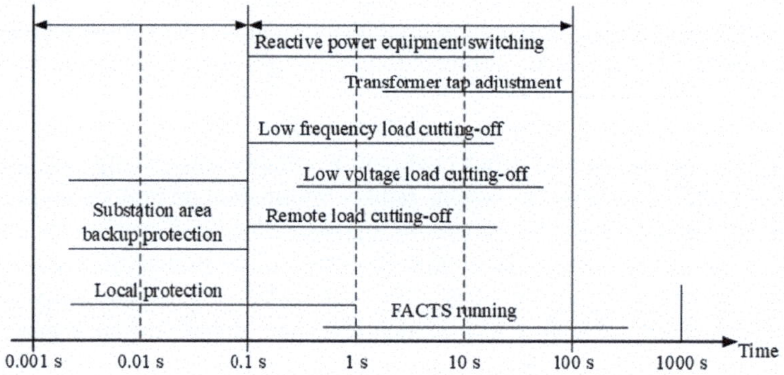

Fig. 5.12 Protection management system functions across various time ranges

are compromised. Global zone protection management relies on fiber-optic communication to ensure data exchange between stations, with reliability often dependent on the fiber-optic network's stability.

Control Questions and Tasks

1. Identify the main functions of a digital substation.
2. What are the goals of substation high-voltage equipment digitalization?
3. How does the architecture of a traditional substation differ from that of a digital substation?

4. Name the primary advantages of digital substations compared to traditional ones.
5. Explain how intelligent condition monitoring of primary substation equipment is performed (choose examples of equipment).
6. Identify major threats to digital substations and methods to mitigate them.

References

1. Krieg T, Finn J, et al (2019) Substations. In: Krieg T, Finn J (eds). Springer International Publishing, p 1079
2. Qi B, Yuan Y, Yang Y, Bu, Q et al (2019) Overview of smart substations. IEC 61850-based smart substations. Academic Press, pp 1–24
3. Zhang C, Cui Y, Ji J, Guo X, et al (2019) Principles and test technology of electronic transformers. IEC 61850-based smart substations. Academic Press, pp 63–89
4. Khalsmaa AI, Yaroshenko SA, Yumanova IF, Stepanova AI, et al (2023) Intelligent multi-agent systems in power engineering. Novosibirsk State Technical University, p 222 (in Russian)
5. Levin VM, Khalsmaa AI, Akhyev DS, Manusov VZ (2020) Application of artificial intelligence methods in technical diagnostics of electrical equipment in power systems. Novosibirsk State Technical University, p 446 (in Russian)
6. Ushakov VYa, Mytnikov AV, Lavrinovich VA, Lavrinovich AV (2022) Transformer condition control advanced and traditional technologies. Springer, p 150
7. Ushakov V, Mytnikov A, Rakhmonov I (2023) High-voltage equipment of power systems. Springer Verlag, p 298
8. Huang Q, Yao R, Li F, Zhen W (2013) Design and implementation of a non-contact magnetic field measurement based fault location system for overhead transmission line. In: IEEE International conference on smart instrumentation, measurement and applications (ICSIMA), pp 1–4
9. Zhang HY, Xie YZ, Yi TQ, Kong X et al (2020) Fault detection for high-voltage circuit breakers based on time-frequency analysis of switching transient E-fields. IEEE Trans Instrum Meas 69(4):1620–1631

Chapter 6
Digitalization of Electric Power Grid Within IEC 61850 Standard

6.1 Content and Scope of the IEC 61850 Standard

All processes of information exchange and control of individual elements at DS are conducted digitally based on the universal IEC 61850 standard, which, with the release of the second edition, was titled "Communication Networks and Systems for Power Utility Automation".

6.1.1 Scope of the IEC 61850 Standard

Many experts refer to IEC 61850 as merely another communication protocol. This is hardly accurate, as a protocol is a set of rules and procedures that enable connection and data exchange between two or more networked devices, while IEC 61850 is a document that includes descriptions of the following fundamental aspects [1]:

– **Information Model**: According to the IEC 61850 standard, the functions of primary and secondary equipment, as well as the equipment itself, have standardized names and information structures. The standardization of data naming and storage structure is critically important, as it enables compatibility in data exchange between devices from different manufacturers, as well as the organization of information exchange between devices using software from independent developers.

  Communication Services: The standard abstractly describes communication models and data transmission services along with their parameters. All of these services are well known and include remote measurements, remote signalling, remote control, event logs, and much more, including communication services for real-time data transmission (fast messaging, instantaneous transmission of current and voltage values). The abstract description of communication models and services ensures their independence from the data transmission technologies

V. Y. Ushakov et al., *Digitalization of Electrical Power Engineering*, Power Systems,
https://doi.org/10.1007/978-3-031-95705-5_6

used, eliminating the need for revisions in light of technological advancements, including data transmission protocols. An analogy to the abstract description of communication services can be found in the literature on the operating principles of, for example, maximum current protection, which is independent of the hardware components used. Only a few chapters of the standard describe the assignment of communication services to a specific communication protocol. Standardizing communication models and services, as well as their assignment to specific communication protocols, is essential to ensure functional compatibility between devices and software from various manufacturers that support the IEC 61850 standard.
- **Universal Configuration Language**: The third component is the universal language for system configuration description, i.e., the description of the aforementioned data models and information exchange between intelligent electronic devices (IEDs). The standardization of this language is another step toward functional compatibility, particularly between software from different manufacturers.

Furthermore, the standard also defines methodologies for verifying the compliance of device and software implementations with the aspects mentioned above.

Considering the fundamental aspects of the IEC 61850 standard, it can be said that functional compatibility between various components of an energy facility's automation system is achieved, but not their interchangeability. The IEC 61850 standard does not inherently provide for component interchangeability by default, but it allows for it within specific facilities and elements.

6.1.2 Overview of the IEC 61850 Standard

The main chapters of the IEC 61850 standard were published between 2002 and 2005. The initial goal of the standard's development was the standardization of information exchange within substation automation systems [2]. In the first edition, the focus was primarily on relay protection and automation (RPA) systems and substation process control systems (PCS). After 2009, these chapters were updated, and new documents were developed to address issues such as organizing measurements, power quality control, equipment monitoring, diagnostics, and more.

As a result, the scope of the IEC 61850 standard expanded beyond substations. Today, the standard covers the following aspects (but, as will be shown, is not limited to them):

- Modelling functions and data of hydroelectric power plants and small-scale generation (IEC 61850-7-410 and 7-420, respectively).
- Modelling functions and data of wind power plants (IEC 61400-25).
- Describing communications between substations, as well as between substations and control centres (IEC 61850-90-1 and IEC 61850-90-2, respectively).
- Modelling functions and data for distribution network automation.

IEC 61850 standard series and associated extensions

IEC/TR 61850-1 ed. 2.0	IEC 61850-7-410 ed. 2.0	IEC/TR 61850-90-1 ed. 1.0
IEC/TS 61850-2 ed. 1.0	IEC 61850-7-420 ed. 1.0	IEC/TR 61850-90-2 ed. 1.0
IEC 61850-3 ed. 2.0	IEC/TR 61850-7-500 ed. 1.0	IEC/TR 61850-90-3 ed. 1.0
IEC 61850-4 ed. 2.0	IEC/TR 61850-7-510 ed. 1.0	IEC/TR 61850-90-4 ed. 1.0
IEC 61850-5 ed. 2.0	IEC 61850-8-1 ed. 2.0	IEC/TR 61850-90-5 ed. 1.0
IEC 61850-6 ed. 2.0	IEC 61850-9-2 ed. 2.0	IEC/TR 61850-90-6 ed. 1.0
IEC 61850-7-1 ed. 2.0	IEC/IEEE 61850-9-3:2016	IEC/TR 61850-90-7 ed. 1.0
IEC 61850-7-2 ed. 2.0	IEC 61850-10 ed. 2.0	IEC/TR 61850-90-8 ed. 1.0
IEC 61850-7-3 ed. 2.0	IEC/TS 61850-80-1 ed. 2.0	IEC/TR 61850-90-10 ed. 1.0
IEC 61850-7-4 ed. 2.0	IEC/TS 61850-80-3 ed. 1.0	IEC/TR 61850-90-12 ed. 1.0
IEC 61850-7-7 ed. 2.0	IEC/TS 61850-80-4 ed. 1.0	IEC/TR 61850-90-17 ed. 1.0

Fig. 6.1 Published chapters of the IEC 61850 standard

Given the above, the current version of the standard has been renamed "Communication Networks and Systems for Power Utility Automation", replacing the older "Communication Networks and Systems in Substations", which reflected its previous limitation to substation systems.

The first edition consisted of 14 documents. Later, some chapters were revised and expanded, and new documents were added to the standard. The current edition consists of 33 documents (Fig. 6.1).

The first chapter of the standard, IEC 61850-1 "Introduction and Overview", was published as a technical report and serves as an introduction to the IEC 61850 series of standards. It defines the basic principles underlying the automation system that operates in accordance with IEC 61850. In particular, it establishes the three-level architecture of the automation system, which includes the process level, bay level, and station level. Initially, the standard defined information exchange within a single energy facility, and connections between facilities, as well as between facilities and control centres, were not considered. The information exchange architecture described in the second edition of the standard, published in 2013, includes a description of these information links.

The second chapter of the standard, IEC 61850-2 "Glossary", was released as technical requirements and contains a glossary of terms, abbreviations, and acronyms used in the context of energy facility automation in the IEC 61850 series of standards.

IEC 61850-3 "General Requirements" was published as a standard and is the only chapter in the series that defines requirements for physical equipment. Among these requirements are descriptions of electromagnetic compatibility of devices, acceptable operating conditions, reliability, etc. This document exists in its second edition, which was published in 2013.

IEC 61850-4 "System and Project Management" was published as a standard and defines all entities involved in the implementation of an energy facility automation system, along with the distribution of responsibilities between them. The following participants are established: the customer (in the form of an electric power company), the design organization or designer, the installation and commissioning organization, and the equipment and software tool manufacturer. The first edition of IEC 61850-4 was published in 2002. Today, the document exists in its second edition, published in 2011.

IEC 61850-5 "Communication Requirements for Functions and Device Models" was published as a standard and defines the conceptual principles of dividing the automation system into levels. It also provides a detailed description of the concept of using logical nodes and proposes their classification according to functional purpose. Additionally, the chapter provides examples of interaction schemes for various logical nodes when implementing a number of protection and control functions (RPA). The document also standardizes the allowable time delays for various types of signals, including discrete signals, sampled instantaneous current and voltage values, time synchronization signals, etc. The document exists in its second edition, published in 2013.

IEC 61850-6 "Configuration Description Language for Communication in Electrical Substations Related to IEDs" was published as a standard and presents a file format for describing the configurations of energy facilities and devices involved in data exchange according to IEC 61850. The main purpose of this file format is to enable device configuration using external software. This file format is known as the Substation Configuration Language (SCL) and is based on the eXtensible Markup Language (XML). The second edition of this chapter of the IEC 61850 standard was published in 2009.

As mentioned above, the IEC 61850 standard also defines approaches for data modelling of real-world objects and abstract communication services. These topics are covered in IEC 61850-7-1, 7-2, 7-3, and 7-4.

In **IEC 61850-7-1** "Basic Communication Structure—Principles and Models", published as a standard, basic methods for system and data modelling are introduced. It presents the principles of data transmission organization and the information models used in other chapters of IEC 61850-7. The document describes the principle of representing a physical device with all its functions as a set of logical devices, which in turn consist of a set of logical nodes. It provides an overview of data transmission principles implemented via "client–server" or "publisher-subscriber" technologies. The document exists in its second edition from 2011.

IEC 61850-7-2 "Basic Information and Communication Structure—Abstract Communication Service Interface (ACSI)", published as a standard, describes the so-called "abstract communication interface" for power utility automation systems. This chapter abstractly addresses services such as remote measurement and remote signalling transmission (so-called reports), remote control services, event logs, file transfers, and more. The document exists in its second edition, published in 2010.

IEC 61850-7-3 "Basic Communication Structure—Common Data Classes", published as a standard, describes all possible data classes and data attribute classes used in IEC 61850 to describe information about status, measured values, control signals, discrete parameters, continuous parameters, and descriptive data. These classes allow for modelling all kinds of data within the substation automation system for subsequent data exchange between devices and systems. The second edition of this chapter was published in 2010.

In **IEC 61850-7-4** "Basic Communication Structure—Compatible Logical Node Classes and Data Object Classes", published as a standard, the information model of devices and functions related to the substation is described. Specifically, it defines the names of logical nodes and data for data transmission between devices, as well as the relationship between logical nodes and data. The document exists in its second edition, published in 2010.

In **IEC Technical Requirements 61850-7-7** "Machine-processable Format of IEC 61850-related Data Models for Tools", the data models of the IEC 61850 standard (logical nodes, common data classes, and other elements) in XML format for processing by various software tools are discussed. These requirements were published in 2018.

IEC 61850-7-410 "Basic Communication Structure—Hydroelectric Power Plants—Communication for Monitoring and Control", published in 2012, describes logical nodes and additional data classes for hydroelectric power plants.

IEC 61850-7-420 "Basic Communication Structure—Distributed Energy Resources and Distribution Automation Logical Nodes", firstly published in 2009, describes logical nodes and additional data classes for distributed energy resources.

IEC Technical Report 61850-7-500 "Basic Information and Communication Structure—Use of Logical Nodes for Modelling Application Functions and Related Concepts and Guidelines for Substations", firstly published in 2012, provides guidelines for using logical nodes defined in IEC 61850-7-4 to model automation functions at substations.

IEC Technical Report 61850-7-510 "Basic Communication Structure—Hydroelectric Power Plants, Steam and Gas Turbines—Modelling Concepts and Guidelines", also firstly published in 2012, provides guidelines for using logical nodes defined in IEC 61850-7-410 to model complex control functions at hydroelectric power plants.

IEC 61850-8-1 "Specific Communication Service Mapping (SCSM)—Mappings to MMS (ISO 9506-1 and ISO 9506-2) and to ISO/IEC 8802-3", published as a standard, examines methods for exchanging information over local networks by mapping the abstract communication services described in IEC 61850-7-2 to the MMS protocol and ISO/IEC 8802-3 (Ethernet) frames. The MMS protocol is used for client–server communications (services for integrating intelligent electronic devices into process control systems), while the Ethernet protocol is used for the Generic Object-Oriented Substation Event (GOOSE) communication service. The first edition of IEC 61850-8-1 was published in 2004, and the second in 2011.

IEC 61850-9-2 "Specific Communication Service Mapping (SCSM)—Sampled Values over ISO/IEC 8802-3", published as a standard, is dedicated to methods for transmitting sampled values from current and voltage transformers (CTs and VTs) via the IEC 8802-3 interface. It defines the mapping of the sampled value transmission class from measuring CTs and VTs of IEC 61850-7-2 onto the ISO/IEC 8802-3 (Ethernet) protocol. The first edition of the IEC 61850-9-2 standard was published in 2004. Today, the document exists in its second edition, published in 2011.

IEC/IEEE 61850-9-3 "Precision Time Protocol Profile for Power Utility Automation", published as a standard, provides a description of the PTP time synchronization protocol profile (according to IEC 61588:2009 and IEEE Std 1588-2008). The profile is designed for use in power utility automation systems and ensures the required precision time synchronization classes according to IEC 61850-5 and IEC 61869-9. The latter is a standard for the digital interface of digital current and voltage transformers. This chapter was published in 2016.

IEC 61850-10 "Conformance Testing", published as a standard, contains descriptions of standard methodologies for verifying the conformance of IEC 61850 implementations in devices (clients and servers) as well as in software (configuration tools for intelligent electronic devices and engineering systems according to IEC 61850). The first edition of IEC 61850-10 was published in 2005, and the second edition in 2012.

IEC 61850-80-1 "Guideline to Exchanging Information from a CDC-based Data Model using IEC 60870-5-101 or IEC 60870-5-104", published as technical requirements in 2016, aims to describe the standard mapping of IEC 61850 data models and communication services onto the data blocks and services of IEC 60870-5-104 or IEC 60870-5-101. The second edition of these technical requirements takes into account the requirements of the second edition of IEC 61850, particularly the requirements of IEC 61850-7-3:2010, which introduces new common data classes.

IEC 61850-80-3 "Mapping to Web Protocols—Requirements and Technical Choices", published as a technical report in 2015, describes the requirements and provides an overview of technical solutions for using web protocols as a new mapping for the abstract communication services and models of the IEC 61850 standard to a specific communication protocol.

IEC 61850-80-4 "Translation from the COSEM Object Model (IEC 62056) to the IEC 61850 Data Model", published as technical requirements in 2016, covers the COSEM specification, which reflects the interface model of metering devices and provides representation of their functional capabilities. The interface model uses an object-oriented approach. Thus, these technical requirements define the mapping between OBIS codes (Object Identification System) of IEC 62056 and the logical nodes of the IEC 61850 standard.

IEC Technical Report 61850-90-1, published in 2010, addresses aspects that should be considered when using the IEC 61850 standard for organizing information exchange between substations, specifically:

- Scenarios where information exchange between substations is required;
- Requirements for this type of information exchange;
- Guidelines for using IEC 61850 communication services and implementing communication architectures;
- Data that must be exchanged as a necessary component for creating functionally compatible solutions;
- The use and extensions of the SCL language to facilitate data exchange between substations.

IEC Technical Report 61850-90-2, published in 2016, examines factors that should be considered when using the IEC 61850 standard for organizing information exchange between substations and control centres. In terms of structure and content, this technical report is similar to the IEC 61850-90-1 report discussed above.

IEC Technical Report 61850-90-3, published in 2016, addresses the organization of monitoring and diagnostic systems for electrical equipment (such as power transformers, cable and overhead power lines, gas-insulated switchgear, etc.) using the IEC 61850 standard.

IEC Technical Report 61850-90-4, published in 2013, focuses on organizing local area networks (LANs) for energy facilities. It examines the advantages and disadvantages of various LAN topologies, information exchange redundancy, time synchronization, limitations on uncontrolled traffic propagation, and other issues according to the IEC 61850 communication services, allowing the designer to make informed decisions. The report also discusses improvements in communication equipment, particularly the implementation of the IEC 61850 data model in Ethernet switches.

IEC Technical Report 61850-90-5, published in 2012, focuses on information exchange between disturbance recorders, phasor measurement units (PMUs), and control centres. The document describes the transmission of data defined in the IEEE C37.118-2005 standard according to the provisions of IEC 61850. It also discusses profiles for routing GOOSE and Sampled Values messages, which can also be used for transmitting phasor measurements.

IEC Technical Report 61850-90-6, published in 2018, covers the fundamental aspects that should be considered when using the IEC 61850 standard to organize information exchange between components and subsystems for implementing automation functions in distribution networks.

IEC Technical Report 61850-90-7, published in 2013, contains a description of the data model for power converters used in small-scale generation, particularly for DC-to-AC and AC-to-AC converters with different voltage levels, photovoltaic systems, energy storage systems, etc., enabling information exchange between distributed generation facilities and control centres.

IEC Technical Report 61850-90-8, published in 2016, explains how the IEC 61850-7-420 standard can be used to model electric transportation systems and their corresponding charging infrastructure.

IEC Technical Report 61850-90-10, published in 2017, describes data models that can be used to model equipment schedules and operating modes. For example, a schedule may be defined by several setpoints corresponding to a specific mode of operation.

IEC Technical Report 61850-90-12, published in 2015, provides guidelines and recommendations for developing distributed communication networks, particularly for organizing relay protection systems, automation, control, and monitoring. The document addresses information exchange between substations, between substations and control centres, and between different control centres. Specifically, it discusses the transmission of protection, automation, and control signals using GOOSE and

Sampled Values communication services, as well as phasor measurements in multicast mode. It also covers the organization of communication infrastructure topology, redundancy, data transmission delays, quality of service, traffic management, time synchronization, security, and other issues.

IEC Technical Report 61850-90-17, published in 2017, focuses on methods of information exchange between power quality measurement devices and corresponding client systems. The primary goal is to ensure functional compatibility between devices in this category. This document addresses the following aspects:

- Guidelines for using the IEC 61850 standard in the field of power quality monitoring;
- The profile for using IEC 61850 in the context of IEC 61000-4-30 "Power Quality Measurement Methods".

In conclusion to the overview of IEC 61850 chapters, the following chapters are recommended for study by specialists in relay protection and automation services:

- **IEC 61850-1**, which describes the overall concept of the IEC 61850 standard.
- **IEC 61850-5**, which provides information on the fundamentals of data modelling according to IEC 61850 and the requirements for the speed of information exchange when solving various tasks (transmission of discrete signals, transmission of sampled values from CTs and VTs, time synchronization requirements, etc.).
- **IEC 61850-6**, which contains information about the system configuration description language, various file formats in SCL, and scenarios for their use when designing energy facilities.
- **IEC 61850-7-1**, which provides information on approaches to modelling functions, including specific examples.
- **IEC 61850-7-2**, one of the key chapters of the standard, describing communication services, including the GOOSE communication service, the sampled values transmission service (Sampled Values), the service for changing values and setting groups, and other services.
- **IEC 61850-7-4**, which contains information on describing relay protection and automation functions in the form of logical nodes and their data. Given that certain electric utility companies have developed corporate profiles based on the IEC 61850 standard, this chapter is key to understanding them.

6.2 IEC 61850-Based Information Model

This section addresses the following topics:

- The main elements of the information model according to the IEC 61850 standard;
- Common data classes defined by the standard;
- The procedure for data modelling;
- Specific features of using elements of the information model.

As noted earlier, IEC 61850 is not just another data transmission protocol. Most chapters of the standard cover conceptual issues of information modelling and data transmission, followed by the so-called abstract communication interface, and finally, the mapping of abstract models and services onto specific communication protocols. This section focuses on the study of information modelling issues according to the IEC 61850 standard. The standardization of data structures and their naming plays a significant role, as it ensures functional compatibility between software or hardware-software solutions from different manufacturers.

6.2.1 Main Elements of the Information Model

The key element of the information model, according to the IEC 61850 standard, is the logical node, which is the smallest unit capable of exchanging data. It can represent one of the constituent functions of a secondary intelligent electronic device (IED). For example, in accordance with Fig. 6.2, the distance protection step is represented by the logical node PDIS, and the overcurrent protection step by the logical node PTOC, and so on [3]. In Fig. 6.2, the logical nodes with a black background represent secondary equipment functions. Additionally, they can also represent physical devices. For instance, a circuit breaker, according to the standard, is represented by the logical node XCBR, while any switching device that does not have arc-quenching capability (such as a disconnector, earthing switch, or load switch) is represented by the logical node XSWI (shown with a white background in Fig. 6.2). A list of all logical nodes described in the standard is presented in the IEC 61850-7-4 chapter. The name of the logical node class always consists of four characters.

In Fig. 6.3, an example of the implementation of an information model in an intelligent electronic device is shown. The physical devices of the main electrical circuit, as well as the protection, control, measurement functions, and other functions, are represented as a virtual model within the intelligent electronic device. In practice, logical nodes can be implemented in one or multiple intelligent electronic devices. Depending on their functional capabilities, logical nodes contain a specific list of data (e.g., position, in the case of a switchgear model, or measurements by a specific phase, in the case of a measurement function model). This data has a specific structure and a clearly defined semantic meaning (i.e., designation in the context of substation automation systems).

An intelligent electronic device can contain several instances of the same logical node. This is necessary, for example, to model several similar switching devices or multiple protection steps within an intelligent electronic device, or different types of protection described by the same class. For instance, if a device controls multiple disconnectors (represented by the XSWI node), each disconnector can be assigned a separate logical node with its own instance number. Similarly, if the device implements maximum overcurrent protection, reverse sequence current protection, and zero-sequence current protection (all described by the PTOC class), each stage of

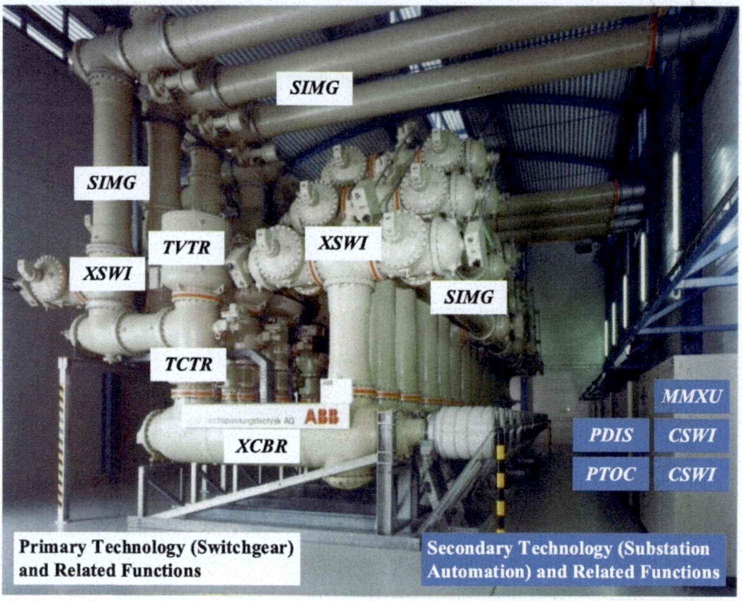

Fig. 6.2 Representation of a logical node at a substation

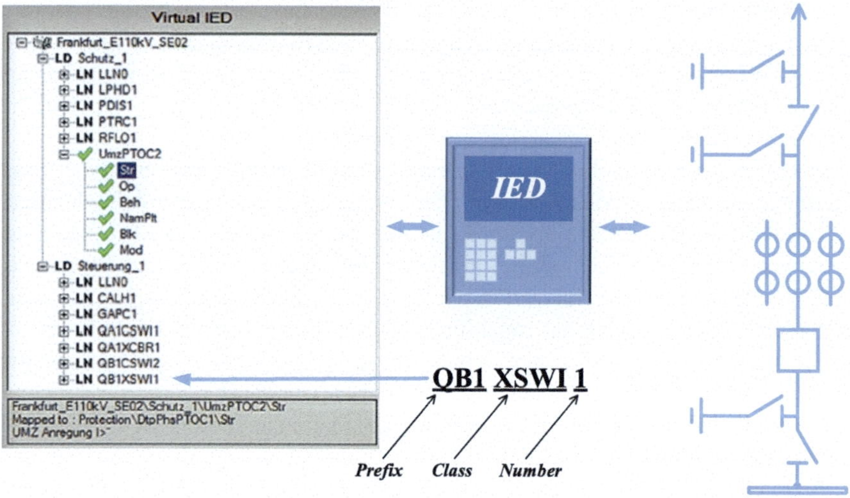

Fig. 6.3 Example of the implementation of the information model in an intelligent electronic device

these functions is assigned a separate logical node. Additionally, for user convenience, a logical node can have a prefix indicating its association with a particular function. As a result, the name of the logical node consists of three parts: the prefix, the class name, and the instance number. In the data model of an intelligent electronic device, no two logical nodes of the same class can have the same instance number.

A logical node occupies a specific position in the data hierarchy [4]. On one hand, it is part of the so-called logical device, implemented in the physical device. On the other hand, as noted earlier, the logical node acts as a data container: within the logical node, there are data objects, which in turn contain data attributes. A logical node can contain from a few to 30 data objects. These data objects can hold from several to 20 (or even more) data attributes. Ultimately, logical nodes can include more than 100 individual pieces of information, organized into a hierarchical structure, as shown in Fig. 6.4.

It is important to note that each data object belongs to a specific data class (or, in standard terminology, the Common Data Class (CDC)). The common data class defines the set of data attributes, each of which has a specific type. The attribute type can be simple, but there are also complex data types, where the attribute itself

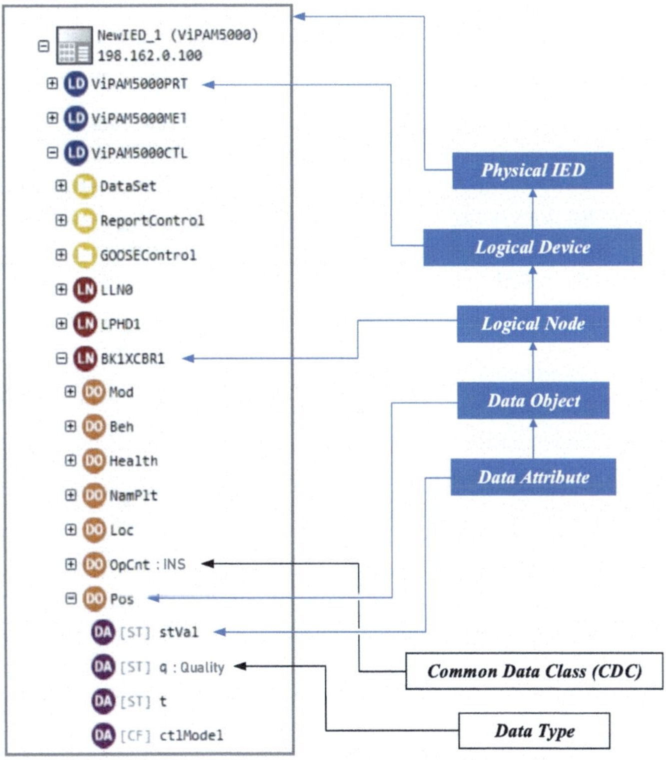

Fig. 6.4 Hierarchical data model

assumes a certain data structure. For example, in Fig. 6.4, there is a simple attribute type—ctlVal (current status), and a complex attribute—q (signal quality), which, in turn, consists of a set of simple attribute types (e.g., BOOLEAN).

Most logical nodes provide information that can be categorized as follows:

- **General information about the logical node**. Information that does not depend on the type of function implemented by the logical node (operating mode, serviceability, passport data, etc.).
- **Discrete information**. Information about the status of the process or function modeled by the logical node (e.g., the activation or triggering of protection systems).
- **Setpoints**. Information necessary for the operation of the function modeled by the logical node, such as the time delay for automatic re-closing (ARC) and other settings.
- **Measurements**. Analog values that are either obtained by direct measurement or computed by various functions (e.g., active, reactive, and total power).
- Controllable data points. Signals that change their value through telecontrol (e.g., switching devices control, on-load tap changers, etc.).

6.2.2 Logical Node and Logical Device

The highest element in the hierarchy of the data model is the logical device, which consists of logical nodes. One or more logical devices may be implemented within a physical device (server). The primary function of logical devices is to group logical nodes. The standard does not specify the number of logical devices within a physical device. This is at the discretion of the manufacturer or the electric grid company as part of the corporate profile of the IEC 61850 standard. However, there are several typical solutions. In devices of protection and automation systems from various manufacturers, logical nodes are grouped in logical devices based on their function: "Protection", "Control", "Measurements" "System", etc. This solution allows, for example, one of the logical devices to be switched into a test mode (TEST mode), where all logical nodes within it are placed in test mode without being activated.

The structure of a logical node will be considered in more detail using the example of the XSWI logical node, which represents a switching device without arc extinguishing capability. Its structure, shown in Fig. 6.5, corresponds to the description of the logical node in the IEC 61850 standard. This logical node contains a set of data objects. For example, the data object "Pos" falls into the category of objects whose value can be controlled. Additionally, in the data model of the XSWI logical node, there is a data object "BlkOpn", which also belongs to the category of controllable data. It is important to understand that each data object has a specific class that, in turn, defines the structure of the data attributes. The standard provides a textual explanation for each data object, unambiguously determining its meaning. Additionally, the standard specifies whether a particular data object is mandatory for implementation by the manufacturer or not [5], where "M" denotes mandatory, "O" denotes optional,

XSWI			
Name	*Class*	*Notation*	
Mod	*ENC*	*Mode*	*C*
Loc	*SPS*	*Local control behaviour*	*M*
EEHealth	*ENS*	*External equipment health*	*O*
EEName	*DPL*	*External equipment name plate*	*O*
OpCnt	*INS*	*Operation counter*	*O*
Pos	*DPC*	*Switch position*	*M*
BlkOpn	*SPC*	*Block opening*	*M*
BlkCls	*SPC*	*Block closing*	*M*
SwTyp	*ENS*	*Switch type*	*O*

- *Mandatory (M) – mandatory data objects*
- *Optional (O) – optional data objects*
- *Condition (C) – mandatory data objects under certain conditions*

Fig. 6.5 Structure of a logical node

and "C" denotes mandatory if the device supports the corresponding communication services that enable operations on the data object in question.

6.2.3 Data Objects and Attributes

The data object "Pos" is more than just a simple data point in the context of supervisory control protocols. It consists of several data attributes (Fig. 6.6), which can be divided into the following categories:

- Status information (or measured/computed values).
- Configuration parameters and descriptive data.
- Substitution and blocking data.

Pos = Position (DPC – Double Point Control)

	Attribute	*Type*	
Status information	*origin*	*Originator*	*Intermediate state | off | on | malfunction*
	ctlNum	*INT8U*	
	stVal	*CODED ENUM*	
	q	*Quality*	***FC = ST***
	t	*TimeStamp*	
	stSeld	*BOOLEAN*	
Configuration parameters and descriptions	*pulseConfig*	*PulseConfig*	*For example, direct control, select before operate (SBO) control*
	ctlModel	*CtlModels*	
	sboTimeout	*INT32U*	
	operTimeout	*INT32U*	
	d	*VISIBLE STRING255*	***FC = CF, DC***
Substitution of values and blocks	*subVal*	*CODED ENUM*	***FC = SV, BL***
	subEna, subQ, subID		
	blkEna		

Fig. 6.6 Data objects and attributes

The main characteristic of the "Pos" object is the stVal attribute (status signal), which can represent four states: "Intermediate Position", "Disconnected" "Connected", and "Fault". These states are typically represented by two bits, and this information is referred to as two-position information. The full set of attributes for the "Pos" object, as mentioned earlier, is defined by the Common Data Class (CDC). The name of the common data class for two-position information is DPC (Double Point Control).

In addition to position, the "Pos" data object contains supplementary information. For instance, it includes the source of the last control command (origin attribute), the command number in the control cycle (ctlNum attribute), or an indication that the control object has already been selected by an external client (stSeld attribute). Other important attributes include q (quality or reliability of the information) and t (timestamp of the last state change of the switching device).

Among the configuration parameters, noteworthy examples include the control pulse configuration "pulseConfig attribute", which may be a single pulse, a sequence of pulses, etc., and the control model "ctlModel attribute", which defines the control method, used "direct control, select-before-execute control", etc.

In the substitution data category, there are attributes such as "subVal" and "subQ", which define the substitute values for the state of the switching device and the reliability of this signal.

As seen in Fig. 6.6, each attribute has a data type—either primitive (e.g., BOOLEAN or INT8U) or complex, with a specific name (e.g., Quality or PulseConfig).

A complex attribute is referred to in the standard as a Constructed Attribute Class. This class is used for attributes within a Common Data Class, as shown in Fig. 6.7. A constructed attribute, in turn, consists of primitive or basic-type attributes (e.g., INT32 or FLOAT32). Through the constructed attribute class, the standard describes information such as data reliability (with detailed possible reasons for unreliability), analog values (for modeling integer or floating-point values), ranges for monitoring (to differentiate various ranges), and other data.

The following sections of the IEC 61850 standard describe logical node classes, data objects, common data classes, and constructed attribute classes.

IEC 61850-7-4 describes:

(1) Logical node classes, including:

- The name of the class and node semantics.
- The composition and names of the data objects.

(2) Data objects, including their semantics and type (with a reference to the common data class).

IEC 61850-7-3 describes:

– (3) Common data classes, including:

- The composition and names of the attributes.

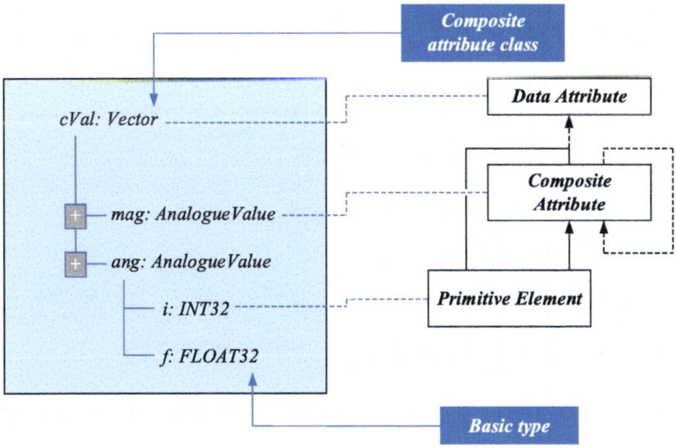

Fig. 6.7 Structure of data attribute

- The semantics and types of the data attributes.

– (4) Constructed attribute classes, including:

- The composition and names of the sub-attributes.
- The semantics and types of the sub-attributes.

IEC 61850-7-2 describes the basic types and common types in accordance with the abstract interface [6].

Figure 6.8 shows the groups of logical nodes designated in the first edition of the IEC 61850 standard. Each group contains a certain number of logical nodes (the number is indicated in parentheses). The letter identifying the group is the first letter in the designation of the logical node classes for that group.

- ***L*** *System Logical Nodes (2)*
- ***P*** *Protection Functions (28)*
- ***R*** *Protection Related Functions (10)*
- ***C*** *Supervisory Control (5)*
- ***G*** *Generic References (3)*
- ***I*** *Interfacing and Archiving (4)*
- ***A*** *Automatic Control (4)*
- ***M*** *Metering and Measurement (8)*
- ***S*** *Sensors (4)*
- ***X*** *Switchgear (2)*
- ***T*** *Instrument Transformer (2)*
- ***Y*** *Power Transformer (4)*
- ***Z*** *Further Power System Equipment (15)*

Examples:

PDIF:	*Differential Protection*	***CSWI:***	*Switch Controller*
RBRF:	*Breaker Failure*	***MMXU:***	*Measuring*
XCBR:	*Circuit Breaker*	***YPTR:***	*Power Transformer*

Fig. 6.8 Logical node groups in the first edition of the IEC 61850 standard

- ***L*** *System Logical Nodes (9)*
- ***P*** *Protection Functions (30)*
- ***R*** *Protection Related Functions (11)*
- ***C*** *Supervisory Control (6)*
- ***G*** *Generic References (4)*
- ***I*** *Interfacing and Archiving (6)*
- ***A*** *Automatic Control (5)*
- ***F*** ***Functional Blocks (9)***
- ***Y*** *Power Transformer (4)*
- ***Z*** *Further Power System Equipment (18)*
- ***K*** ***Mechanical and Non-electric Primary Equipment (5)***
- ***M*** *Metering and Measurement (13)*
- ***Q*** ***Power Quality Events (6)***
- ***S*** ***Supervision and Monitoring (11)***
- ***T*** ***Instrument Transformers and Sensors (20)***
- ***X*** *Switchgear (2)*

Modifications:

- ***New groups of logical nodes***
- ***New logical nodes***
- ***Extensions to new applications (hydroelectric power plants, small-scale generation, etc.)***

Fig. 6.9 Logical node groups in the second edition of the IEC 61850 standard

For example, in group P (Protection), according to the first edition of the IEC 61850 standard, there are 28 logical nodes. One of them (PDIF) is a differential protection logical node. Meanwhile, group X (Switching Devices) contains 2 logical nodes, one of which (XCBR) represents a circuit breaker, and so on.

In the second edition of the standard, shown in Fig. 6.9, the number of logical node groups and the number of logical nodes within existing groups were increased (partly due to the expanded scope of IEC 61850). For instance, new logical node groups were added for primary mechanical equipment (such as ventilation and pump equipment), for power quality and level monitoring devices (instrument transformers and sensors), and for various functional blocks.

The standard does not provide a complete list of logical nodes for all possible tasks and functions. However, it offers clearly defined rules for describing additional logical nodes and data, for instance, for additional functions within substations or for other applications. Missing logical node classes can be defined by customers, as is done by the European Network of Transmission System Operators for Electricity (ENTSO-E) in Europe [7].

The following is a closer look at the description of the circuit breaker logical node, which is provided in IEC 61850-7-4. Firstly, the following semantics are defined: "This logical node is used for modelling switching devices with arc-quenching capability. Other logical nodes may be required for comprehensive modelling of the device (e.g., SIMS, which controls the insulation medium—oil, SF6, etc.). Operating commands for switching on and off should be generated by the CSWI (switching device control node) and CPOW (controlled switching node) logical nodes (if applicable). If the control service with time delay is not implemented, switching commands should be generated using the GSE service (see IEC 61850-7-2)".

Secondly, the data objects of the XCBR logical node and their textual explanations are defined, for instance, the "Pos" object (Position), the "BlkOpn" object (Open Block), the "BlkCls" object (Close Block), and so on, as shown in Fig. 6.10.

For each data object, it is specified whether it is mandatory for implementation by the manufacturer (M), optional (O), or mandatory if the device supports the

XCBR Class Logical Node			
Data Object Name	**Common Data Class**	**Explanation**	**M/O/C**
Description			
EEName	DPL	External equipment name plate	O
Status information			
EEHealth	ENS	External equipment health	O
LocKey	SPS	Local or remote key (local means without substation automation communication, hardwired direct control)	O
OpCnt	INS	Operation counter	M
CBOpCap	INS	Circuit breaker operating capability	O
POWCap	INS	Point On Wave switching capability	O
MaxOpCap	INS	Circuit breaker operating capability when fully charged	O
Dsc	SPS	Discrepancy	O
Metered values			
SumSwARs	BCR	Sum of switched Amperes, resetable	O
Controls			
LocSta	SPC	Switching authority at station level	O
Pos	DPC	Switch position	M
BlkOpn	SPC	Block opening	M

Fig. 6.10 Structure of the XCBR logical node

appropriate communication services that allow operations on the object in question. It is also shown that each data object corresponds to a specific Common Data Class (CDC).

Thirdly, the semantics of all data objects are described. For example, the data object CBOpCap is defined as an enumeration of the physical capabilities of the circuit breaker. This object reflects the number of Close-Open cycles as well as the lockout status due to malfunctions. The value of the CBOpCap object is always less than or equal to the value of the MaxOpCap object (the maximum number of Close-Open cycles under full readiness).

The next point is to take a closer look at the Common Data Classes (CDC), as defined by the IEC 61850 standard. In the context of the IEC 61850 standard, a CDC significantly reduces the complexity of data definitions. When referring to data, it is not necessary to list all the attributes of the data; it is enough to indicate the CDC. CDCs are also an excellent tool for unifying the structure of data attributes. The IEC 61850-7-3 chapter defines CDCs for a wide range of applications. The main data classes are grouped into the following categories:

- Discrete information (10 classes)
- Measurements (9 classes)
- Control (8 classes)
- Discrete set points (7 classes)
- Analog set points (3 classes)
- Descriptive information (3 classes).

Figure 6.11 provides an example of the SPS (Single Point Status) CDC, as described in IEC 61850-7-3.

The data in this class is divided into the following categories:

Single Point Status (SPS) class					
Attribute Name	Type	FC	TrgOp	Value / Value Range	M/O/C
Status					
stVal	BOOLEAN	ST	dchg	TRUE | FALSE	M
q	Quality	ST	dchg		M
t	TimeStamp	ST			M
Substitution					
subEna	BOOLEAN	SV			PICS_SUBST
subVal	BOOLEAN	SV		TRUE (ИСТИНА) | FALSE (ЛОЖЬ)	PICS_SUBST
subQ	Quality	SV			PICS_SUBST
subID	VISIBLE STRING64	SV			PICS_SUBST
blkEna	BOOLEAN	BL			O
Configuration, Description and Extension					
d	VISIBLE STRING255	DC		Text	O
dU	UNICODE STRING255	DC			O
cdcNs	VISIBLE STRING255	EX			AC_DLNDA_M
cdcName	VISIBLE STRING255	EX			AC_DLNDA_M
dataNs	VISIBLE STRING255	EX			AC_DLN_M

Fig. 6.11 Example of SPS CDC

- Status information (or measured/computed values);
- Configuration parameters and descriptive data;
- Substitution and blocking data.

Each category of the class defines a set of attributes. For each attribute, the data type is specified, as well as the Functional Constraint (FC), whose purpose will be discussed later; the trigger type that causes sporadic transmission of information (dchg—on signal value change, qchg—on signal quality change); the range of possible values; and the manufacturer's implementation condition (mandatory/optional).

The IEC 61850-7-3 chapter also describes the semantics for each data attribute. As an example, one could consider the semantics for the timestamp (t) of the last state change of one of the data attributes or the quality (q) of the information. For different CDCs, the timestamp corresponds to changes in the state of the respective attributes: for the SPS CDC, it is stVal, for the DPS CDC, it is also stVal, and so on, where stVal represents a single-point status signal.

Some CDCs contain data objects that reference another CDC. One example of such a CDC is WYE, which represents measurements in a three-phase system. Figure 6.12 shows that in a separate group of this class, sub-objects are highlighted, whose type is the CMV class (complex measurement).

The complex values describe vectors. This is a good example of a constructed attribute class—Vector, as shown in Fig. 6.13, which itself consists of other constructed attributes: mag and ang—the magnitude and phase of the complex value, which in turn consist of basic-type data attributes.

When examining data class descriptions, you will come across the abbreviation FC—Functional Constraint. This parameter defines the characteristic of the information being considered: for example, whether it is status information (ST) or measured/calculated values (MX), and so on. Using functional constraints allows access to the necessary information within a data object without explicitly listing it, for example,

WYE Class					
Attribute Name	Type	FC	TrgOp	Value / Value Range	M/O/C
Data Object					
phsA	CMV				GC_1
phsA	CMV				GC_1
phsA	CMV				GC_1
neut	CMV				GC_1
net	CMV				GC_1
res	CMV				GC_1
Data Attribute					
Configuration, Description and Extension					
angRef	ENUMERATED	CF	dchg	Va | Vb | Vc | Aa | Ab | Ac | Vab | Vbc | Vca | Vother | Aother | Synchrophasor	O
phsToNeut	BOOLEAN	CF	dchg	DEFAULT = FALSE	O
d	VISIBLE STRING255	DC		Text	O
dU	UNICODE STRING255	DC			O
cdcNs	VISIBLE STRING255	EX			AC_DLNDA_M
cdcName	VISIBLE STRING255	EX			AC_DLNDA_M
dataNs	VISIBLE STRING255	EX			AC_DLN_M

Fig. 6.12 Example of the complex WYE CDC

Vector			
Attribute Name	Type	Value / Value Range	M/O/C
mag	AnalogueValue		M
ang	AnalogueValue	$-180 < n \leq +180$	AC_CLC_O

- ***mag:*** *amplitude of complex value*
- ***ang:*** *phase of complex value*

Fig. 6.13 Example of the vector constructed attribute class

during configuration. Figure 6.14 provides an example of functional constraints for a data object of the SPS class.

In order to transmit three signals—status (stVal), quality (q), and timestamp (t)—from an object of this class to other devices, according to a particular communication service of the IEC 61850 standard, it is sufficient to configure the object with the functional constraint FC equal to ST. In this case, other attributes (or in other words, signals) of the object in this class will not participate in the data transmission.

The IEC 61850-7-2 chapter defines the basic data types that are used to describe the data objects of the class:

- BOOLEAN (logical variable).
- INT8/16/32/64 (signed integer with the corresponding number of bits).
- INT8U/16U/24U/32U (unsigned integer with the corresponding number of bits).
- FLOAT32 (32-bit floating point).
- ENUMERATED (enumerated data type).
- CODED ENUM (enumerated data type with predefined values).
- OCTET STRING (byte string).
- VISIBLE STRING (text string).
- UNICODE STRING (Unicode string).

Functional constraints are used to classify attributes based on their purpose:

- ***ST*** *– status information*
- ***MX*** *– measurands (analogue values)*
- ***SP*** *– setting*
- ***SG*** *– setting group*
- ***SE*** *– setting group editable*
- ***DC*** *– description*
- ***CF*** *– configuration*
- ***SV*** *– substitution*
- ***SR*** *– service response*
- ***OR*** *– operate received*
- ***BL*** *– blocking*
- ***EX*** *– extended definition*

SPS Class		
Attribute Name	***Type***	***FC***
Status		
stVal	*BOOLEAN*	*ST*
q	*Quality*	*ST*
t	*TimeStamp*	*ST*
Substitution/Blocking		
subEna	*BOOLEAN*	*SV*
subVal	*BOOLEAN*	*SV*
subQ	*Quality*	*SV*
subID	*VISIBLE STRING64*	*SV*
blkEna	*BOOLEAN*	*BL*
Configuration, Description and Extension		
d	*VISIBLE STRING255*	*DC*
dU	*UNICODE STRING255*	*DC*
cdcNs	*VISIBLE STRING255*	*EX*
cdcName	*VISIBLE STRING255*	*EX*
dataNs	*VISIBLE STRING255*	*EX*

Fig. 6.14 Functional constraints

Summarizing the key elements of the information model of the IEC 61850 standard, the following can be noted: intelligent electronic devices (IEDs) implement a hierarchical data model, the structure of which corresponds to the functions performed by these IEDs (Fig. 6.15).

Within a physical device, one or more logical devices can be implemented, each containing a certain set of logical nodes corresponding to the functions being performed. Typically, logical nodes are grouped into logical devices based on their functional groupings. Logical nodes consist of data objects and data attributes, representing the range of input and output signals of the modelled functions.

Logical nodes contain data objects, which, in turn, consist of data attributes, the list of which is determined by the Common Data Classes (CDC) defined in the standard, particularly in the IEC 61850-7-3 chapter. The data attributes within the data objects can have a basic type or can be constructed, meaning they include a structure of attributes of a basic type. The data objects represent the input and output signals of functions. For example, in the XCBR circuit breaker logical node, this could include the position of the switching device, with the data attributes of this object being the state of the device, the reliability of this signal, and the timestamp of its last change.

6.2.4 Data Modelling Procedure

Data modelling is an important step not only in the development of specific technical solutions by the equipment supplier but also in the development of the IEC 61850 standard profile by the customer [8]. In the first stage of data modelling according to the IEC 61850 standard, the following tasks are addressed:

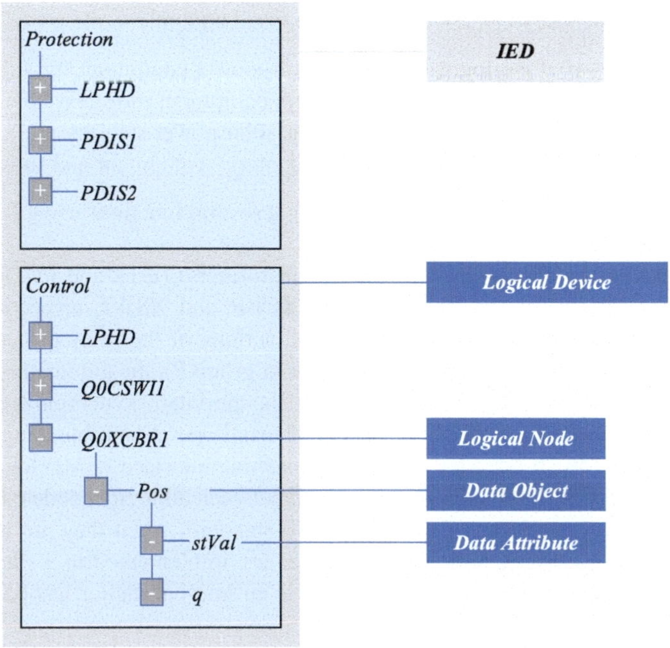

Fig. 6.15 Example of a hierarchical data model in an IED

1. Determining the list of functions used within subsystems:
   - Relay protection and automation.
   - SCADA.
   - Measurement and power quality control, etc.
2. Determining the classes of logical nodes and their quantities:
   - Stepwise distance protection.
   - Circuit breakers, disconnectors, earthing switches, etc.
   - On-load tap changers (OLTC), etc.
3. Clarifying whether the use of logical nodes not defined by the standard is required. In other words, it is necessary to determine whether new logical nodes need to be created.

When determining the correspondence of functions and equipment to the classes of logical nodes according to the IEC 61850 standard, the question arises as to where to find logical nodes already regulated by the standard. Several chapters of the IEC 61850 standard address this issue:

- IEC 61850-7-4: Contains a list of logical nodes:
  - For use in substation automation systems (including relay protection and automation).

- General-purpose (e.g., converters, mechanical equipment, etc.).

- IEC 61850-7-410: Lists logical nodes for hydropower equipment and subsystems.
- IEC 61850-7-420: Describes logical nodes for equipment and subsystems of small generation facilities (e.g., small CHP plants, solar power stations, etc.).
- IEC 61400-25: Lists logical nodes for wind energy equipment and subsystems.

Figure 6.16 provides an example of modelling several functions using IEC 61850 logical nodes.

The logical nodes for current and voltage transformers (TCTR and TVTR, respectively), circuit breakers and disconnectors (XCBR and XSWI, respectively) are shown. Among the functions presented are: functions of the relay protection and automation logical nodes group (logical nodes in group P), the automatic reclosing function (represented by logical node RREC), the automatic synchronization function (RSYN), switching device control (CSWI), and operational blocking (CILO). Additionally, Fig. 6.16 includes the human–machine interface (HMI) logical node (IHMI). Signals and commands are exchanged between all logical nodes, following the communication services of the IEC 61850 standard, when they are located in different physical devices. When logical nodes are implemented in a single physical device, the question of node interaction is an internal matter resolved by the equipment developer.

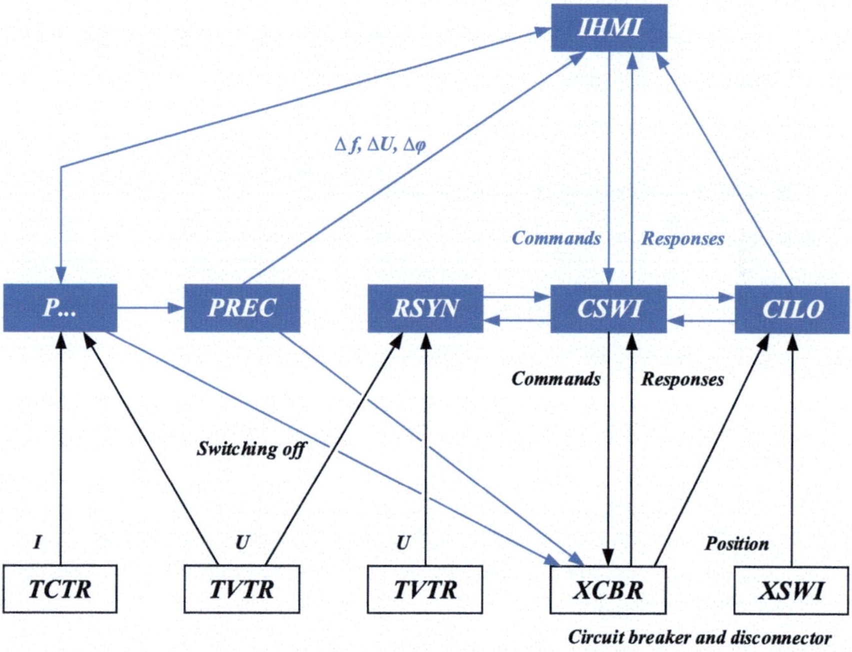

Fig. 6.16 Example of function modelling

In cases where the standard does not provide a suitable logical node to represent the required function, it will be necessary to create a new logical node class in accordance with the standard's requirements.

In the second stage of data modelling, the following tasks are solved:

1. Determine the information required for use within the logical nodes, namely:
   - Data objects,
   - Data attributes. This is an important step since all communications, in accordance with the IEC 61850 standard, involve the transmission of data objects and attributes.
2. Evaluate the need to create new data objects that are not defined by the standard. If necessary, they are created based on the common data classes already described in the standard.

In the third stage of information modelling, the following tasks are performed:

1. Distribute the IEC 61850 logical nodes across physical devices.
2. Distribute the IEC 61850 logical nodes across logical devices.
3. Define the names of the elements of the information model. Logical devices must be assigned specific names, and logical nodes must be assigned specific prefixes and instance numbers that uniquely identify the type of implemented functions. Assigning prefixes and instance numbers to logical nodes may also be carried out in earlier stages of the data model specification.

As a result of information modelling, a data model is created for each physical device in the substation, Fig. 6.17, which includes a certain number of logical devices with defined names, as well as logical nodes with specified prefixes and instance numbers.

In a similar manner, the task of developing an IEC 61850 standard profile in a specific organization and specifying the data model of devices with different functional purposes can be solved. Largely, the typification of the information model for devices supporting the IEC 61850 standard is analogous to the typification of terminal blocks for devices and cabinets of traditional design.

Control questions and Tasks

1. Explain the term "digital substation". What is the generalized structure of a digital substation?
2. Name the main advantages and disadvantages of digital substations. What drives the gradual transition to digital substations?
3. What is covered by the set of IEC 61850 standard chapters? Give examples of the contents of specific chapters of IEC 61850.
4. What is a logical node according to the IEC 61850 standard?
5. Define intelligent electronic devices. How do they differ from traditional protection and automation devices at substations?

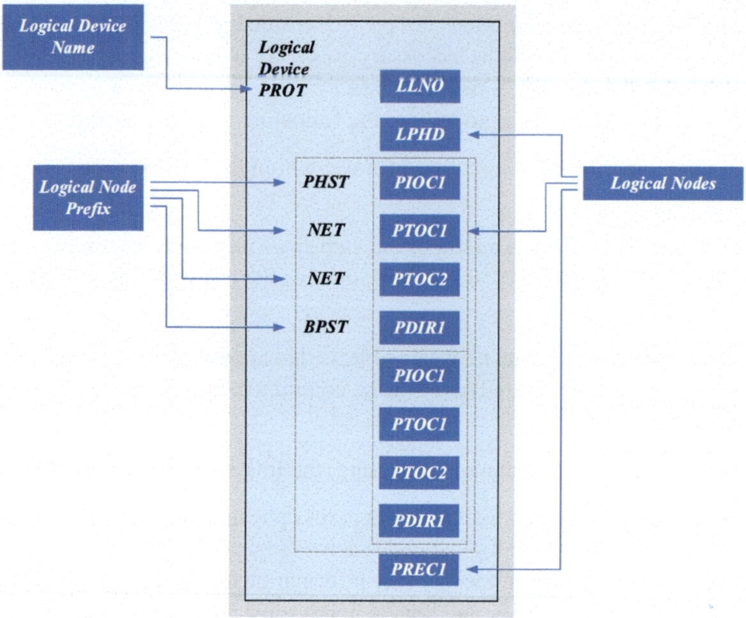

Fig. 6.17 Example of a device data model

6. Explain the purpose of using different data objects and attributes in intelligent electronic devices.
7. What does the hierarchical data model look like in an intelligent electronic device?

References

1. Hughes R, Bishop P, Nair N (2023) Introduction to IEC 61850 systems. IEC 61850 principles and applications to electric power systems. Springer International Publishing, pp 21–42
2. Huang Z, Gao L, Yang Y, Kong X et al (2019) IEC 61850 standards and configuration technology. IEC 61850-based smart substations: principles, testing, operation and maintenance, pp 25–62
3. Aftab M, Hussain S, Ali I, Ustun TS (2020) IEC 61850 based substation automation system: a survey. Int J Electr Power Energy Syst 120:106008
4. Loken R (2023) IEC 61850 communication architectures and services. IEC 61850 principles and applications to electric power systems. Springer International Publishing, pp 67–93
5. MacKiewicz RE (2006) Overview of IEC 61850 and benefits 623–630
6. Kumar S, Abu-Siada A, Das N, Islam S (2023) Review of the legacy and future of IEC 61850 protocols encompassing substation automation system. Electronics 12(15):3345
7. Ayello M, Lopes Y (2023) Interoperability based on IEC 61850 standard: systematic literature review, certification method proposal, and case study. Electr Power Syst Res 220:109355
8. Himri Y, Muyeen S, Malik F, Himri S et al (2021) Review on applications of the standard series IEC 61850 in smart grid applications. Cyberphysical Smart Cities Infrastruct Optim Oper Intell Decis Mak 197–253

Chapter 7
IEC 61850 Requirements for Digital Substation Design

7.1 SCL Configuration Language

The IEC 61850-6 standard defines SCL, a language used for the formal description of devices and core systems of a digital substation, which is based on XML [1]. The SCL language is used to describe the following objects:

- System configuration; intelligent electronic devices (IEDs);
- Functional specification of the system;
- Configured IEDs;
- Connections between switching devices, equipment, and substation control systems.

7.1.1 *Types of SCL Files*

To ensure data exchange compatibility between IED configurators from different manufacturers, the system configurator, and for exchanging developed files (including individual project components), IEC 61850-6 defines appropriate data exchange formats [2]. All electronic descriptions of the substation and related systems are contained in files with different functional purposes and, consequently, different file name extensions. According to the IEC 61850 standard, there are five different scenarios for data exchange using SCL, and therefore, five types of SCL files are distinguished (Fig. 7.1) for data exchange between Computer-Aided Design (CAD) systems.

The distinction between file types is made by using different file extensions, including the following [3]:

- **.SSD File** (System Specification Description)—A system specification description file created by CAD for system specification. The SSD file may optionally describe virtual IEDs, and each virtual device can be matched with a specification

V. Y. Ushakov et al., *Digitalization of Electrical Power Engineering*, Power Systems,
https://doi.org/10.1007/978-3-031-95705-5_7

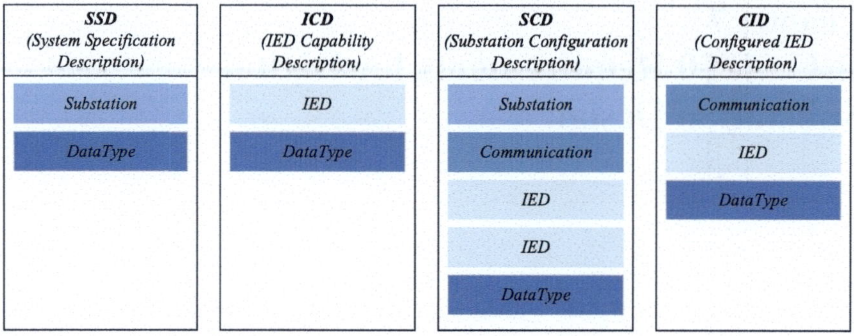

Fig. 7.1 Types of SCL files

in the form of an ISD file (IED Specification Description). Currently, efforts are underway to standardize this file format (it is the only format not yet standardized). This will allow customers to specify devices with different functional purposes according to the IEC 61850 standard.

- **.ICD File** (IED Capabilities Description)—A file describing the functional capabilities of an IED, created by the IED configurator.
- **.SCD File** (System Configuration Description)—A file describing the system configuration, created by CAD for configuring the design object.
- **.CID File** (Configured IED Description)—A file describing the configuration of the IED, created by the IED configurator.
- **.IID File** (Instantiated IED Description)—A file describing the modified configuration of the IED (a specific case of the ICD format file).
- **.SED File** (System Exchange Description)—A file describing the configuration of a neighbouring system (e.g., an adjacent substation).

The SSD file describes the single-line diagram of the power facility and the requirements for relay protection and automation (RPA) functions, control systems, commercial energy metering, etc., for the connections of the design object in the form of logical nodes and their data. This file is intended to transfer data from system specification software to system engineering software (Figs. 7.2 and 7.3).

Fig. 7.2 SSD file structure

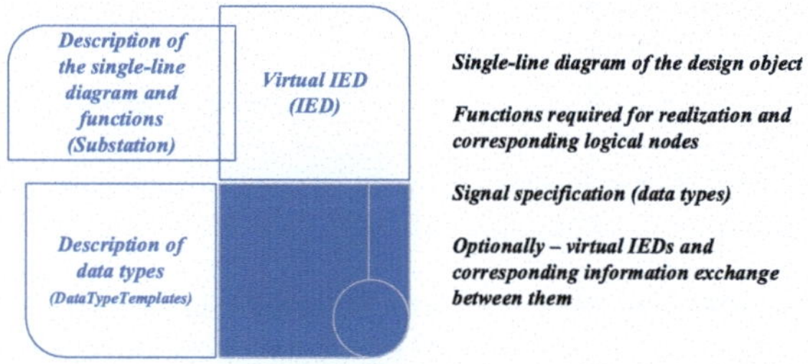

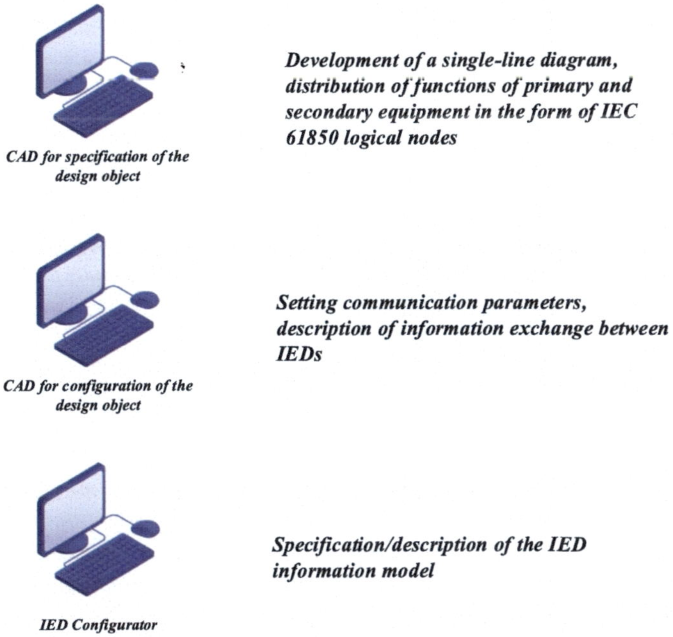

Fig. 7.3 Classification of software tools

From a structural perspective, the SSD file includes the Substation and DataTypeTemplates sections. The Substation section provides a description of the single-line diagram of the design object, as well as the functions of primary equipment and secondary subsystems for the connections of the power facility in the form of logical nodes. The DataTypeTemplates section contains the data structure of logical nodes used to describe the functions of primary equipment and secondary subsystems. The data structure of the logical nodes defines the settings, input and output signals of the represented functions. Optionally, the DS file may contain a description of virtual IEDs with requirements for their data models and communication services.

If the SSD file contains a description of virtual electronic devices, an ISD-format file can be generated from it to describe the virtual electronic devices. These files are intended to transfer the configuration of virtual electronic devices to software used for selecting equipment suppliers during competitive procedures (to assess compliance with IEC 61850 requirements) or to software used for certification to assess compliance with the IEC 61850 profile of the customer.

An ISD file (Fig. 7.4) includes the IED section (which describes the data model of the IED and its communication services) and the DataTypeTemplates section (which describes data templates). Optionally, it may include the Substation section, which describes the design object, and the Communications section, which describes the communication services supported by the IED.

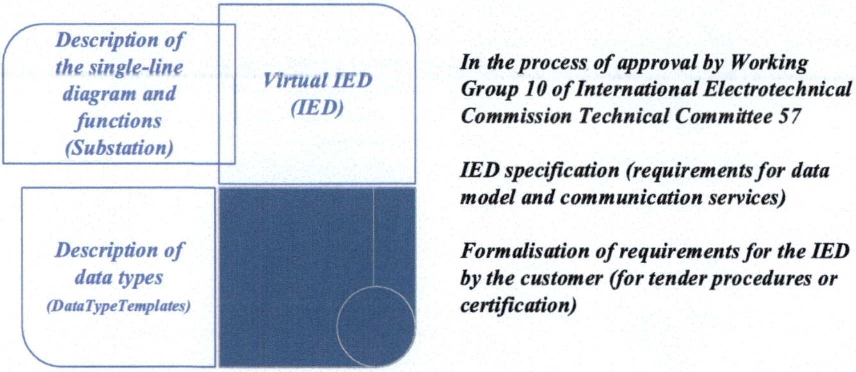

Fig. 7.4 ISD file structure

The ICD file describes the functional, communication, and configuration capabilities of an IED. It includes a description of the IED's information model in the format of logical nodes and their data, a description of the supported communication services of the IEC 61850 standard, and restrictions on their configuration. This file is intended for transferring data from the IED configuration software to the system engineering software.

The ICD file, as shown in Fig. 7.5, includes sections for IED (Intelligent Electronic Device) and DataTypeTemplates. The IED section describes the logical nodes supported by the IED's data model (functions such as protection and automation, process control systems, commercial metering, etc.). The DataTypeTemplates section describes the structure of the logical nodes used to represent the functions of the IED. Optionally, ICD files can include the Substation and Communications sections, depending on the specific device.

The SCD file describes the single-line diagram of the power facility, as well as secondary subsystems such as protection, automation, process control systems,

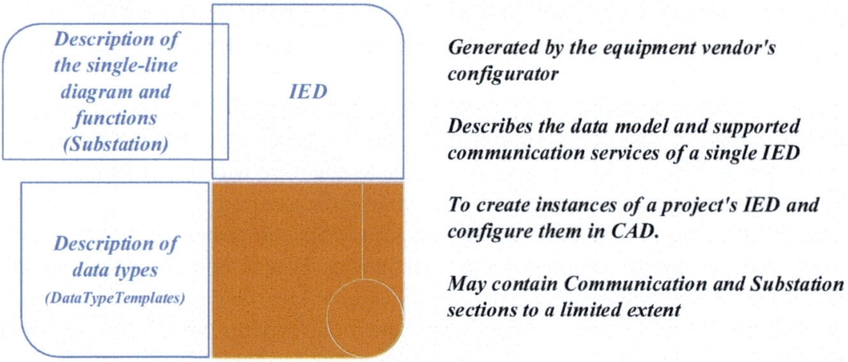

Fig. 7.5 ICD file structure

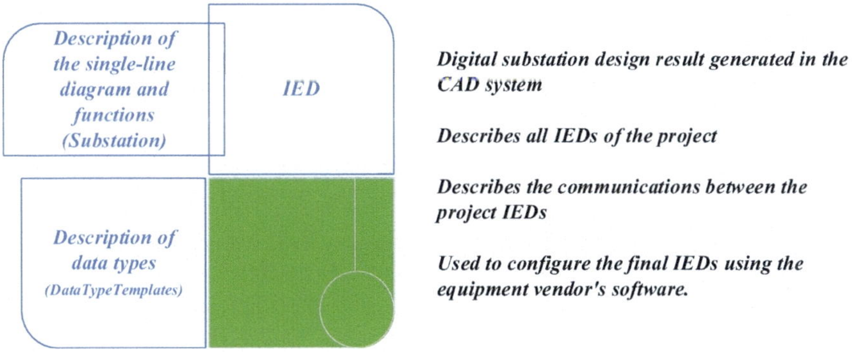

Fig. 7.6 SCD file structure

commercial metering, and others, including all IEDs of the secondary subsystems, the functions they implement in the form of logical nodes associated with the facility's connections, and the configuration of information links between them. This file is intended for transferring data from system engineering software to IED configuration software.

The SCD file (Fig. 7.6) includes sections such as Substation, Communication, IED (one for each device in the project), and DataTypeTemplates.

The Substation section differs from that in the SSD file, as it includes the assignment of primary and secondary equipment functions (logical nodes according to IEC 61850) defined in the SSD file to the corresponding functions (logical nodes) implemented by the selected IEDs. The IED section differs from that in the ICD file with the addition of prefixes and suffixes for the corresponding logical nodes, as required by the SSD file, along with data transfer settings (the set of signals transmitted by each IED, the communication service used for data transmission, data transmission service parameters, etc.). The IED section also defines the data received by the specified IED from external devices via communication services according to the IEC 61850 standard, through the description of external data references in the Inputs section.

The CID file describes the configuration of a specific IED in the project (information model, communication services, communication settings). This file is intended for transferring data from the IED configuration software directly to the IED.

In fact, the CID file is a reduced version of the SCD file. The file may also describe other IEDs in the project with which the current IED interacts, but only in the part that concerns the data received from those IEDs. Thus, the CID file (Fig. 7.7) always includes sections for IED, Communication, and DataTypeTemplates. Optionally, it may include the Substation section, in the part relevant to the current device.

The IID file is a specific case of the ICD file format. This format is used in cases where the number of logical node instances for an IED can be configured by the user based on the specific configuration of the object and the corresponding requirements for the functionality of secondary subsystems. For example, when a system is

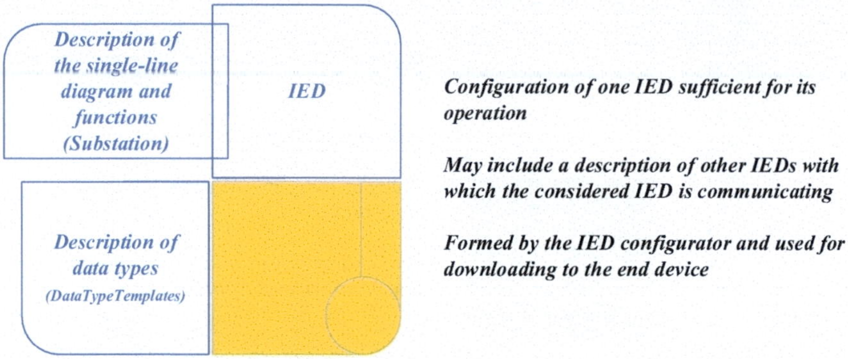

Fig. 7.7 CID file structure

designed to use a feeder controller that can be equipped with different numbers of discrete input/output modules. Depending on the number of installed modules, this controller can monitor and control various switching devices. This means that the information model of this controller will vary depending on the specific use case. Such a data model, tailored to the specific design object and feeder controller, will be expressed in the IID file. This file is intended for transferring data from the IED configuration software to the system engineering software.

The IID file, as shown in Fig. 7.8, includes sections for IED, Communication, and DataTypeTemplates. The IED section describes the logical nodes supported by the IED data model (functions such as protection, automation, process control systems, commercial metering, etc.). The Communication section may contain descriptions of device communication parameters and communication service parameters. The DataTypeTemplates section provides the structure of the logical nodes used to describe the IED functions. Optionally, the IID file may include the Substation section.

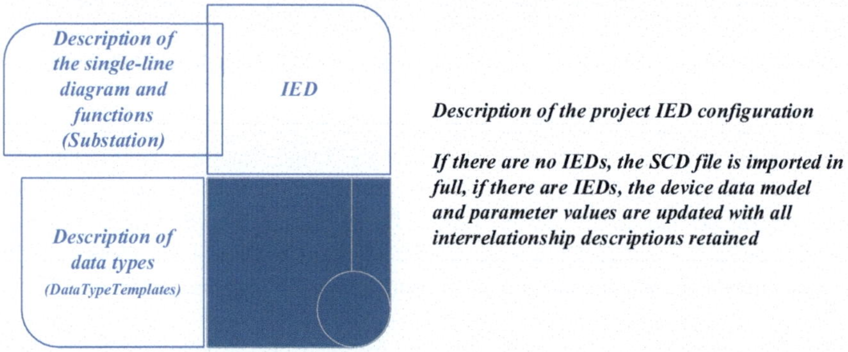

Fig. 7.8 IID file structure

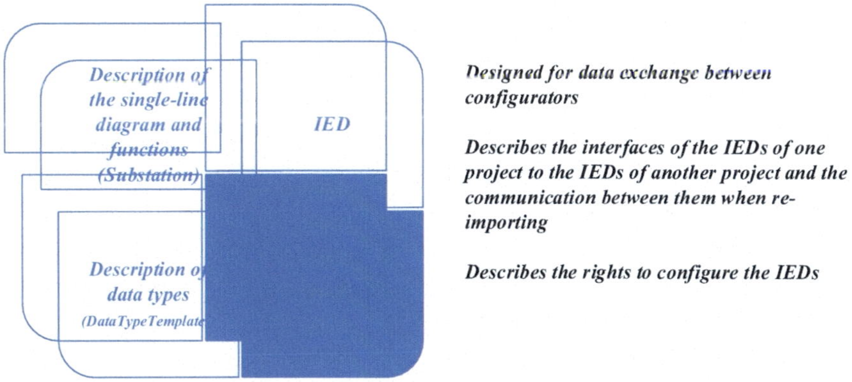

Fig. 7.9 SED file structure

The SED file describes the communication interfaces of the IEDs in one project for use in other projects, with appropriate rights to modify their parameters. Upon re-import into the original project, it also includes the additionally configured interface connections between the IEDs of the projects. This file is intended for data exchange between CAD systems for the configuration of an object.

The file is a subset of the SCD file, containing interface fragments of IEDs to which connections will be established from other projects, thereby preserving the links to the source object already specified by the references. Compared to the SCD file, the SED file defines both the access rights for modifying the configuration of each IED and the main project to which the IED belongs, in terms of project usage (import). The SED file (Fig. 7.9) includes the same sections as the SCD file, but in a limited scope.

It is assumed that all information exchange between participants during the design of a digital substation (DSS) should be carried out using files created in the SCL language. These files accompany the DSS project throughout its entire lifecycle. The information exchange model during the design process using SCL files is shown in Fig. 7.10.

7.1.2 Classification of Software Tools

To create and interpret files in the SCL language, the IEC 61850-4 standard introduces the following types of software tools [4]:

1. **System Specification Tool**

   This software tool allows the formalized description of components such as primary equipment, the single-line diagram of the substation, and an abstract, named set of required functions without specifying the specific IEDs that will

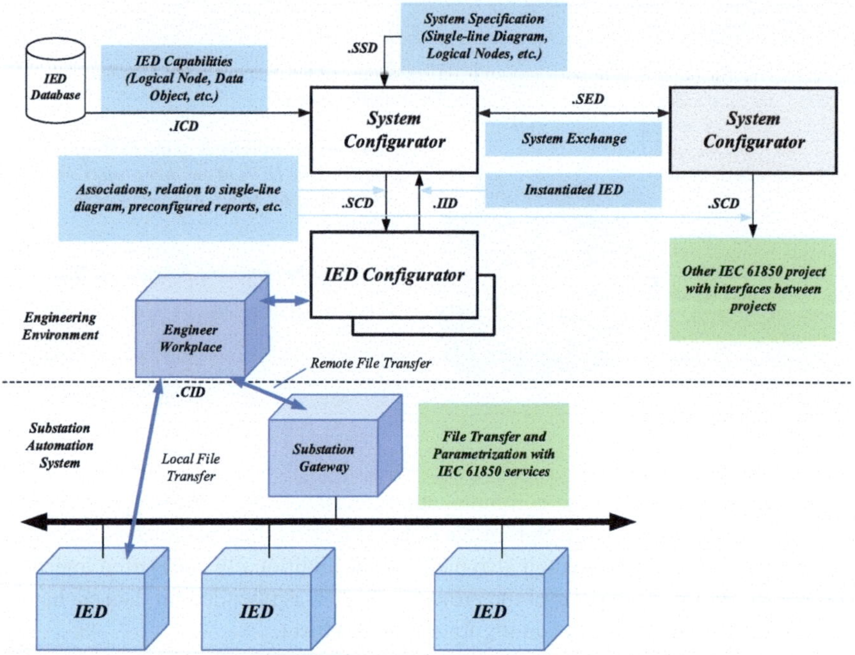

Fig. 7.10 Information exchange model in the design process according to IEC 61850-6

perform them, during the project documentation stage in the SCL language. Typically, this tool uses a database of templates for standard functions and their required signals, as well as typical components of the technological process. The system specification described in an SSD file serves as the input data for subsequent stages of designing the automated system. The system specification tool should be independent of IED manufacturers.

2. **System Configuration Tool**

This software tool allows the selection of components with the necessary functions for the automation system during the working documentation stage. The tool contains a database of IED capability descriptions or ICD files provided by IED manufacturers. It also allows for the configuration of all communications between IEDs required to perform the necessary automation functions of the substation.

The resulting system configuration in the form of an SCD file provides a complete description of the substation automation system according to the IEC 61850 standard. This description, written in the formal syntax of the SCL language, enables the semi-automatic use of system configuration information by other software tools that support the SCL language at all subsequent stages, such as for configuring IEDs and/or creating electronic documentation. The tool should allow importing IED descriptions and interfaces, as well as exporting system configuration descriptions.

The system configuration tool should generally be independent of IED manufacturers. However, it may also provide extended support for configuring IEDs specific to a particular manufacturer, utilizing proprietary elements of the SCL syntax.

3. **IED Configuration Tool**

   This tool is designed to create a specified set of functions and parameters for a particular IED within an automated system. This set of parameters is defined by the IED manufacturer or, more precisely, by the type of IED. The main functions and parameters of the IED, as well as other system data, should be imported from the system configuration description created using the system configuration tool. For this purpose, the IED configuration tool must support importing configuration descriptions from an SCD file in the SCL language. Any subsequent new functions, parameters, and settings for the IED will also be configured using this IED configuration tool.

   The IED configuration tool is typically developed and provided by the IED manufacturer. The relationship between the system configuration tool and the IED configuration tool is schematically shown in Fig. 7.11.

4. **Documentation Tool**

   A system documentation tool generates documentation in the required format based on the formalized description of the system configuration contained in the SCD file, in accordance with relevant standards. The information stored in the SCD file includes descriptions of hardware and software platforms that implement the specified functions, along with all interfaces and connections

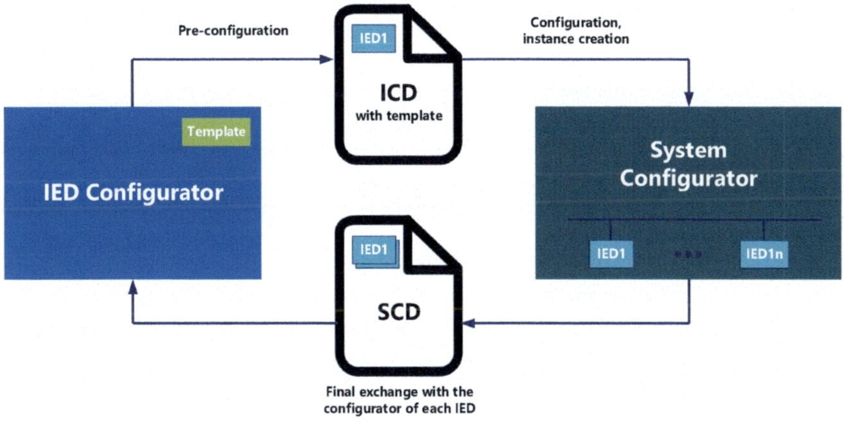

Fig. 7.11 IED description in the system configurator

between IEDs. The file also contains sufficient descriptions of the system's software components to create functional diagrams, block diagrams, and parameter tables, reflecting all connections made during configuration.

The capabilities of the documentation tool should be based on the implementation of descriptions according to IEC 61850-6 and may be supplemented with specific IED properties described in private fields. IEC 61850-4 also does not impose restrictions on combining the functions of different configuration tools in a single software product.

7.2 Digital Substation Design

According to IEC 61850-4 and IEC 61850-6, the implementation of a digital substation (DSS) involves developing electronic project documentation files in the SCL language, as discussed in the previous section. These files are equivalent to the project and working documentation developed for objects created using "traditional" principles. The electronic project documentation files contain the data model of the design object, descriptions of all intelligent electronic devices (IEDs) at the power facility, and the information connections between them, including the communication parameters of these links [5]. Moreover, these electronic project documentation files can be loaded into software for configuring the IEDs during the commissioning stage, allowing the devices to start exchanging data according to the developed project solutions. In the following section, the development of electronic project documentation will be discussed in more detail.

7.2.1 Design Process According to IEC 61850 Standard

The IEC 61850 standard defines the business process for designing digital substations, which is schematically presented in Fig. 7.12.

According to this process, at the initial stage of design, the SSD file (System Specification Description) is developed for the design object. The following tasks are addressed during the creation of the SSD file using the SCL language:

- Description of the primary circuit of the design object;
- Description of the required functions for secondary subsystems such as protection and control, automation, process control systems (PCS), commercial energy metering, and others for the connections of the design object in the form of standard objects from the IEC 61850 information model.

In practice, the development of the SSD file corresponds to the stage of creating a placement scheme for the design object and the adjacent network of RPA, recloser, and automatic load transfer devices during traditional design [6].

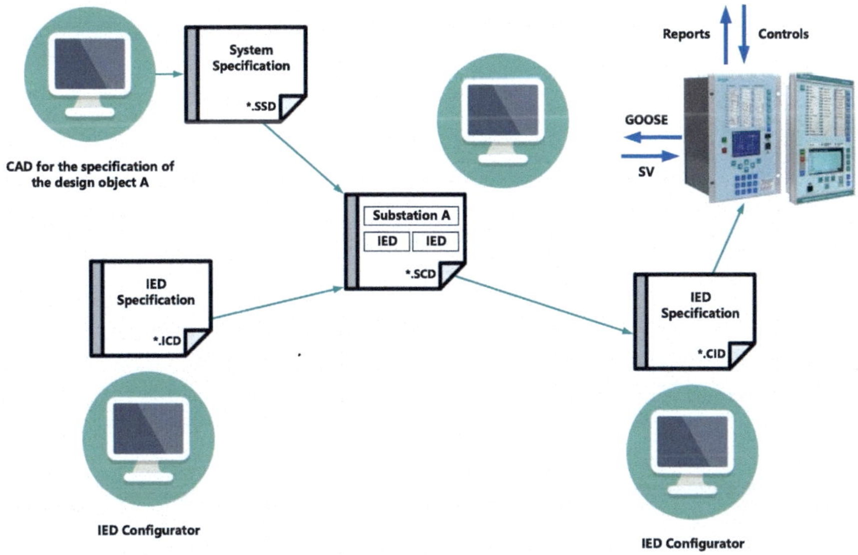

Fig. 7.12 Design process according to IEC 61850

At the same time, manufacturers of RPA devices, PCS, and other IEDs that support the IEC 61850 standard provide descriptions of their supported functions in ICD files (IED Capabilities Description) using the same standard objects from the IEC 61850 information model. This allows for the verification of the proposed technical solutions' compliance with the network company's requirements, as described in the semantics and syntax of IEC 61850, either during the equipment certification stage (when an approved IEC 61850 profile is available) or during competitive procedures, based on the specific SSD file for selecting equipment suppliers for the project.

The SSD file and the ICD files are loaded into the system engineering software, where the IED data model is mapped to the data model in the SSD file, and the communication links between the IEDs are configured. The SCD file describes the allocation of SSD functions (logical nodes) across the IEDs, the communication parameters of the IEDs, the information exchange between the IEDs at the design object according to GOOSE and Sampled Values communication services, as well as the communication services for integrating IEDs into the PCS (reports, event logs, operational control, etc.). The developed SCD file is a part of the working documentation and is intended for loading into software for configuring the IEDs.

In the following sections, the purpose, structure, and development process of SSD, ICD, and SCD electronic project documentation files will be discussed in more detail.

7.2.2 Purpose and Structure of SSD, ICD and SCD Files

Below, the purpose, development process, and structure of the System Specification Description (SSD) file are examined. The SSD file is developed for the following objectives:

- Formulating the requirements of the electric utility company for the representation of secondary system functions such as protection and control (RPA), process control systems (PCS), and their signals in accordance with the designations of the IEC 61850 standard. Since, at the later stage of working documentation development, all information exchange between IEDs is described by configuration files based on the semantic requirements of the SSD file, and the transmitted signals do not have names in the messages transmitted via the local area network of the power facility, the requirements for representing the functions of primary and secondary equipment in the semantics of IEC 61850 are critically important to ensure the subsequent identification of signals and, consequently, for the operation of RPA, PCS, commercial energy metering, and other systems.
- The SSD file formalizes the requirements for IED functionality in terms of the set of logical nodes according to IEC 61850 and can be used within competitive procedures for equipment selection or for equipment certification procedures.
- The SSD file, and in particular, the formalized description of the primary circuit of the design object in the SCL syntax, allows for the automation of mnemonic frame creation for process control systems at subsequent stages of project implementation.

To represent the functions of secondary systems and primary equipment in a formalized manner, the IEC 61850 standard defines the concept of logical nodes, as shown in Fig. 7.13.

For example, a specific class of logical node can represent the function of overcurrent protection (PTOC), breaker failure protection (RBRF), reclosing (RREC), measurements (MMXU), and others. Each class of logical node has a standardized data structure, which can include settings, input, and output signals. Although the data structure of the logical node is defined by the standard, not all are mandatory for implementation by every manufacturer, and in some cases, logical nodes do not contain the signals that are typically used in the secondary systems of certain network companies. Therefore, the task of describing the required functions of secondary systems involves not only identifying the list of logical nodes from the IEC 61850 standard corresponding to the required functions but also determining the internal data structure of the logical nodes that the project intends to use.

Additionally, prefixes and instance numbers can be assigned to logical nodes to distinguish between similar functions (for example, overcurrent protection, negative-sequence protection, zero-sequence protection, etc.), which are represented by the same logical node class according to IEC 61850 (e.g., PTOC). For instance, prefixes allow the differentiation of various functions represented by the same logical node class, while instance numbers can specify the number of instances of functions

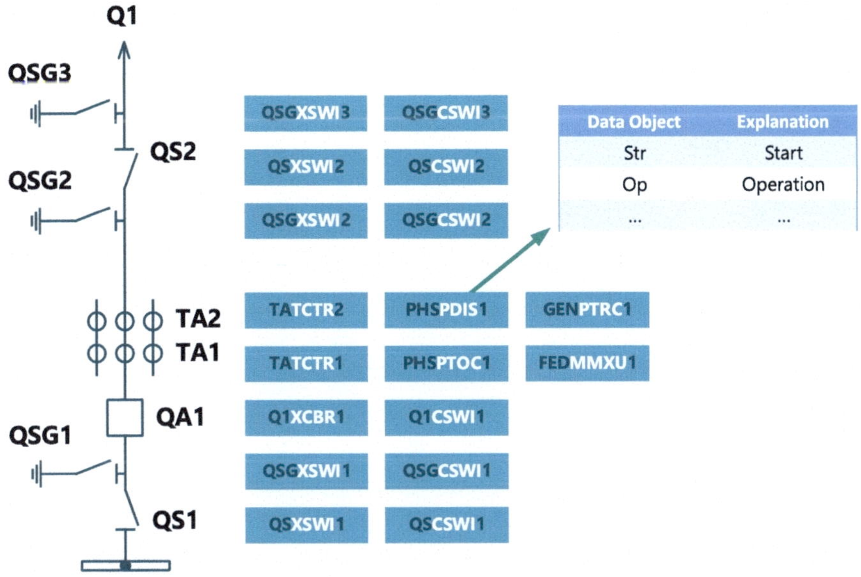

Fig. 7.13 SSD file development process

required to be implemented in the IED (e.g., the required number of stages of RPA functions).

The set of required functions to be implemented at the design object is determined by the applicable regulatory documents of the electric utility company. The correspondence of logical nodes to secondary system functions, the naming of logical nodes (prefixes and instance numbers), and the data structure of logical nodes are defined by the profiles of the IEC 61850 standard created by electric utility companies.

In accordance with the IEC 61850 standard, the SSD file includes the Substation section (Fig. 7.14), which describes the primary circuit of the object and the distribution of functions represented in the form of logical nodes from the IEC 61850 standard across the equipment elements in the Substation section.

The SSD file also includes the DataTypeTemplates section (Fig. 7.15), which describes the types of logical nodes, data objects, and data attributes used.

It is worth noting that SSD files are developed using specialized software that does not require knowledge of the SCL language and allows working in an intuitive graphical environment. It includes a library of logical nodes according to the IEC 61850 standard and/or the electric utility company's profile.

Figure 7.16 shows a graphical interpretation of the SSD file. Based on the presented scheme, it can be seen that the development of the SSD file corresponds to the stage of creating a placement scheme at the design object and in the adjacent network of RPA, recloser, automatic transfer switch, PA, and other devices.

The purpose of the ICD file is as follows:

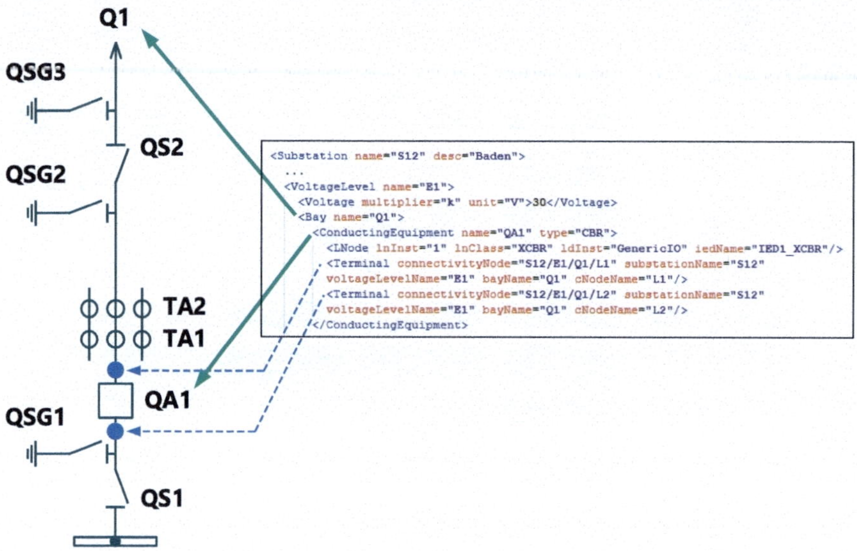

Fig. 7.14 Substation section of the SSD file

```
<DataTypeTemplates>
  ...
  <LNodeType id="IED_000000001/CTRL/Q0XCBR1" lnClass="XCBR">
    <DO name="Mod" type="myMod_11"/>
    <DO name="Beh" type="myBeh_4"/>
    <DO name="Health" type="myHealth_5"/>
    <DO name="NamPlt" type="myNamPlt_18"/>
    <DO name="Loc" type="myProxy_10"/>
    <DO name="OpCnt" type="myFltNum_184"/>
    <DO name="Pos" type="myPos_231"/>
    <DO name="BlkOpn" type="mySPCSO1_218"/>
    <DO name="BlkCls" type="mySPCSO1_218"/>
    <DO name="CBOpCap" type="myCBOpCap_234"/>
  </LNodeType>
```

Fig. 7.15 DataTypeTemplates section of the SSD file

- The ICD file corresponds to a specific modification of the IED and is developed and provided by the equipment supplier.
- The ICD file describes the logical nodes supported by the IED's data model (protection functions, process control systems, commercial energy metering, etc.) as well as the structure of the logical nodes' data used to describe the IED's functions.
- The ICD file also describes the supported communication services of the IED (types of communication services supported, permissible number of messages for sending and receiving, etc.).

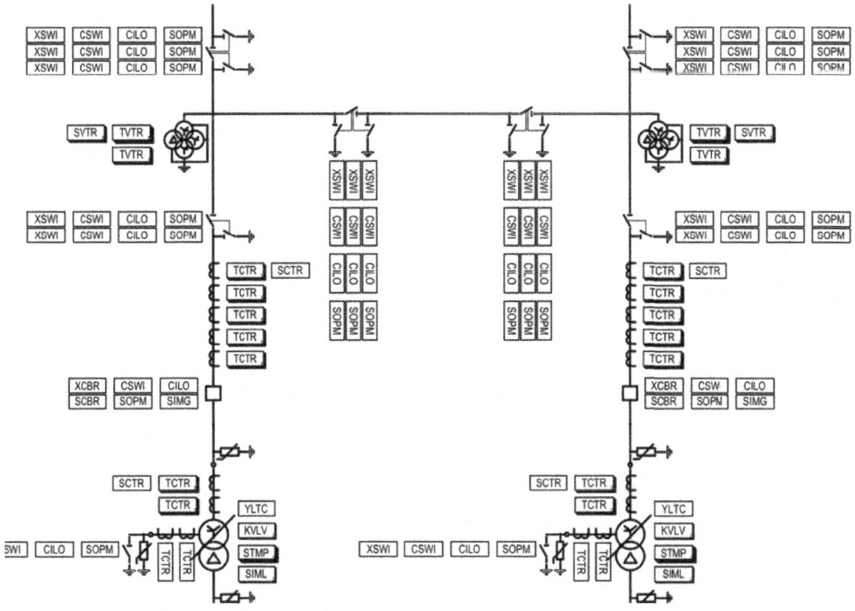

Fig. 7.16 Graphical interpretation of the SSD file

The ICD file consists of two main sections—IED and DataTypeTemplates. The IED section contains several subsections, the main ones being Server and Services. The DataTypeTemplates section is similar in structure to that in the SSD file, which was discussed earlier in Fig. 7.16.

The Server subsection of the IED section describes the data model of the device, specifically, the list of functions of primary equipment and secondary systems (protection and control, process control, commercial metering, etc.) supported by logical nodes. Each logical node belongs to a specific data type template. A fragment of the IED section in SCL syntax is shown in Fig. 7.17.

The Services subsection of the IED section describes the GOOSE, Sampled Values, Report, and other communication services supported by the selected IED configuration, as well as the restrictions on the configuration of these communication services (such as the number of digital messages transmitted, the ability to configure certain parameters of the supported communication services, etc.). This section is used by the system configuration software to identify the IED's functional limitations and prevent the project specialist from implementing invalid technical solutions. A fragment of the Services subsection in SCL syntax is shown in Fig. 7.18.

The SCD file describes:

- The primary circuit of the design object.
- All intelligent electronic devices (IEDs) in the project, including their data model and communication parameters within the project.

```
<IED name="IED1_XCBR">
  ...
  <AccessPoint name="accessPoint1">
    <Server>
      <Authentication />
      <LDevice inst="GenericIO">
        <LN0 lnClass="LLN0" lnType="LLN01" inst="">
          ...
        </LN0>
        <LN lnClass="LPHD" lnType="LPHD1" inst="1" prefix="" />
        <LN lnClass="XCBR" lnType="XCBR1" inst="1" prefix="">
          ...
        </LN>
        <LN lnClass="CSWI" lnType="CSWIa" inst="1" prefix="">
          ...
        </LN>
        ...
      </LDevice>
    </Server>
  </AccessPoint>
</IED>
```

Fig. 7.17 IED section of the ICD file

```
<IED name="IED1_XCBR">
  <Services>
    <DynAssociation />
    <GetDirectory />
    <GetDataObjectDefinition />
    <GetDataSetValue />
    <DataSetDirectory />
    <ReadWrite />
    <GetCBValues />
    <ConfLNs fixPrefix="true" fixLnInst="true" />
    <GOOSE max="5" />
    <GSSE max="5" />
    <FileHandling />
    <GSEDir />
    <TimerActivatedControl />
  </Services>
```

Fig. 7.18 Services subsection of the ICD file

- The information connections between the IEDs of the project, indicating the set of signals transmitted/received by each device, the communication services used for information exchange, and the communication parameters for each exchange.
- It is used to load the project solution into IEDs during commissioning at the power facility. The SCD file includes the following sections: Substation, Communication, IED, and DataTypeTemplates.

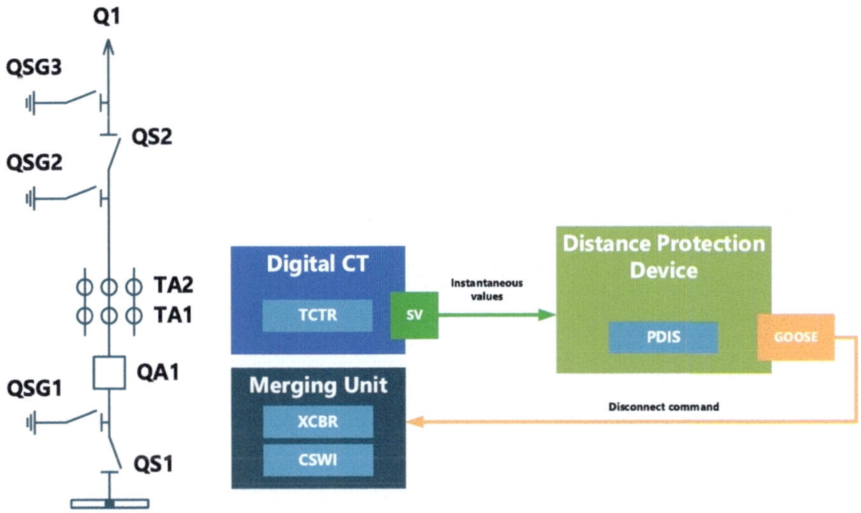

Fig. 7.19 Example of mapping to real equipment in the SCD file

The Substation section differs from that in the SSD file by linking the functions of primary and secondary equipment (logical nodes according to IEC 61850), defined by the SSD file, to the corresponding functions (logical nodes) implemented by the IEDs selected for the project. For example, Fig. 7.19 shows the placement of the current transformer logical node TCTR in the digital current transformer, the distance protection logical node in the distance protection device, and the circuit breaker and switching control logical nodes in the merging unit, which is intended to collect discrete signals from the circuit breaker, issue control signals, and send and receive information to/from the IED using the GOOSE communication service. Figure 7.19 also shows the information exchange between the project devices.

To implement the mapping in the developed SCD file, which includes the description of the design object's scheme with the required functionality in the form of logical nodes (in the Substation section), cross-references are made to the corresponding logical nodes of the relevant IEDs (in the IED sections). A fragment of the SCD file in SCL syntax is shown in Fig. 7.20, which illustrates the mapping of a logical node in the Substation section to the corresponding logical node of the merging unit as part of the project.

The IED section differs from that in the ICD file by having prefixes and instance numbers for the corresponding logical nodes, as required by the SSD file, as well as data transfer settings. Specifically, this section outlines the set of signals transmitted by each IED, indicates the communication service used for data transmission, defines the communication service parameters for data transmission, and more. The IED section also defines data sets for transmission according to GOOSE, Sampled Values, and other communication services of the IEC 61850 standard.

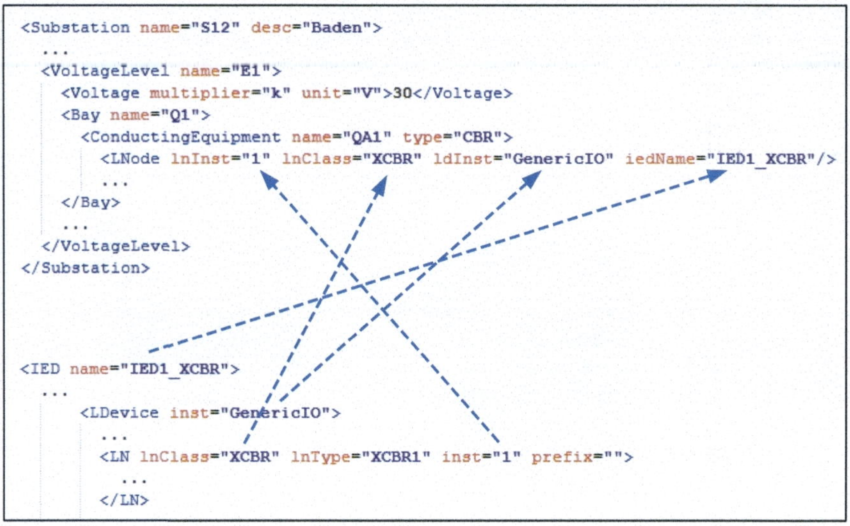

Fig. 7.20 Substation and IED sections of the SCD file

The Communication section provides a description of the communication parameters for the transmission of GOOSE messages, Sampled Values streams, telemetering/tele-signalling, and more. A fragment of the SCD file in SCL syntax is shown in Fig. 7.21, which describes the transmission of a GOOSE message from the distance protection device.

In the IED section, data received by the specified IED from external devices is also defined, according to the communication services of the IEC 61850 standard, through the description of external data references in the Inputs subsection of the IED. A fragment of the SCD file in SCL syntax is shown in Fig. 7.22, where the reception of a GOOSE message from the distance protection device by the merging unit is described.

Ultimately, as a result of the development of the SCD file, each IED in the project knows which functions it implements, what communication parameters it has, what information it sends and receives, and according to which communication protocols. The electronic project documentation in SCD format is uploaded into each IED in the project, and these IEDs begin functioning according to the planned project solutions.

7.2.3 *File Validation*

This section addresses the validation of electronic project documentation files written in SCL, specifically the SSD files and SCD files [7]. When validating SSD files, the specialist must tackle the following tasks:

```xml
<Communication>
  <SubNetwork name="subnetwork1" type="8-MMS">
    ...
    <ConnectedAP iedName="IED2_PTOC" apName="accessPoint2">
      ...
      <GSE ldInst="GenericIO" cbName="gcbEvents">
         <Address>
              <P type="VLAN-ID">1</P>
              <P type="VLAN-PRIORITY">4</P>
              <P type="MAC-Address">01-0c-cd-01-00-02</P>
              <P type="APPID">1002</P>
         </Address>
    </GSE>
    </ConnectedAP>
    ...
</Communication>


<IED name="IED2_PTOC">
  ...
      <LDevice inst="GenericIO">
        <LN0 lnClass="LLN0" lnType="LLN01" inst="">
          <DataSet name="Events" desc="Events">
            <FCDA ldInst="GenericIO" lnClass="GGIO" fc="ST" lnInst="1" doName="SPCSO1" daName="stVal" />
            <FCDA ldInst="GenericIO" lnClass="PTRC" fc="ST" lnInst="1" doName="Tr" daName="general" />
          </DataSet>
          <GSEControl appID="events" name="gcbEvents" type="GOOSE" datSet="Events" confRev="2"  />
        </LN0>
          ...
      </LDevice>
    ...
</IED>
```

Fig. 7.21 Communication subsection of the SCD file

```xml
<IED name="IED1_XCBR">
  ...
    <LDevice inst="GenericIO">
      ...
      <LN lnClass="LPHD" lnType="LPHD1" inst="1" prefix="" />
      <LN lnClass="XCBR" lnType="XCBR1" inst="1" prefix="">
        <Inputs>
          <ExtRef intAddr="Tr" desc="Trip" iedName="IED2_PTOC" ldInst="GenericIO" lnClass="PTRC"
          lnInst="1" doName="Tr" daName="general" serviceType="GOOSE" srcLDInst="GenericIO" srcLNClass="LLN0" srcCBName="GoCB" />
          <ExtRef intAddr="Op" desc="Operate" iedName="IED1_XCBR" ldInst="GenericIO" lnClass="CSWI"
          lnInst="1" doName="Pos" daName="opOk" serviceType="Poll" />
        </Inputs>
      </LN>
      ...
    </LDevice>
  ...
</IED>
```

Fig. 7.22 Inputs subsection of the SCD file

1. Syntax validation and checking for compliance with the SCL schema or, as commonly referred to, its validation. Any discrepancies with the standard's requirements regarding SSD file composition will prevent them from being loaded into system configurators to create the SCD file, as well as into verification software.
2. Validation of semantically significant logical nodes and their correspondence to the implemented protection and control functions. As previously mentioned, each logical node represents a specific function of primary or secondary equipment or the equipment itself. It is crucial that appropriate logical node classes are used in the SSD file following the IEC 61850 standard and, if applicable, the corresponding IEC 61850 profile. For instance, zones of distance protection

should be modelled using logical nodes of the PDIS class, differential protection using the PDIF class, and a circuit breaker using the XCBR class, etc.

In addition to using the required set of logical nodes, it is important to ensure that the data objects and attributes of the logical nodes are sufficient to enable information exchange between devices. If an IEC 61850 profile exists, the corresponding types of logical nodes must be used.

Validation of the SSD file, like any file based on a language derived from XML, is performed relative to a specific schema. Based on this schema, not only can the correctness of the document's syntax be checked, but also its structure, content model, and data types. The schema for the SCL system configuration description language is defined in Chap. 6 of the IEC 61850 standard. With appropriate software tools, compliance checks are carried out automatically. The main requirement under such conditions is the generation of a validation report in a format that can be understood by the verifier and the file developer, indicating the cause of errors.

Validation of the required set of logical nodes ensures that the designer has correctly mapped the protection and control (RPA) functions of individual elements of the electrical network to the IEC 61850 standard logical node classes. Additionally, it is necessary to check that prefixes and instance numbers are assigned to the logical nodes to allow further identification of RPA functions and stages. Each logical node should have a brief description. This task is greatly simplified if there is an approved IEC 61850 profile, which essentially standardizes the set of logical node classes, their prefixes, instance numbers, and textual descriptions for each element of the electrical network.

When validating the SSD file, it is not only necessary to check the set of logical nodes, but also to ensure that these nodes contain the required set of data objects and attributes. The composition of these elements is determined by the logical node type. The importance of this procedure lies in the fact that these data objects and attributes will be transmitted between devices using various communication services of the IEC 61850 standard. Equipment suppliers will be required to support an agreed-upon set of data objects and attributes. Each data object and attribute must have a textual description to allow for unambiguous identification of signals. As with logical nodes, the validation task is significantly simplified when an approved IEC 61850 profile exists, as the profile defines the types of logical nodes as well as the names of all data objects and attributes.

When validating SCD files, the specialist must address the following tasks:

1. Syntax validation and ensuring compliance with the SCL schema, similar to the validation performed for SSD files discussed earlier. This validation is extremely important, as the SCD file is subsequently transferred to the commissioning stage for loading into the device configurators. Any discrepancies with the standard's requirements in the composition of the SCD files will prevent their loading into the device configurators, testing setups (e.g., Omicron, PONOVO with IEC 61850 support), and other software and hardware systems, as these configurators typically perform SCD file validation during import.

2. Validation of functional connections between devices according to GOOSE and Sampled Values communication services. The proper operation of protection and automation systems depends on the correct organization of information exchange via Sampled Values messages (for transmitting instantaneous current and voltage values) and GOOSE messages (discrete signals and analogue values). An analogy in the traditional approach to secondary switching circuits would be the RPA wiring and installation diagrams. Errors in these schematic solutions regarding the connection of elements in the RPA complex (output contacts, intermediate relays and their contacts, discrete inputs, etc.) can lead to incorrect protection and automation operation and must be identified.
3. Verification of communication parameters for GOOSE and Sampled Values messages. Incorrectly set communication parameters may hinder the transmission of Sampled Values and GOOSE messages across the network. An analogy in the traditional approach would be a mismatch between the operating voltage of a device's discrete input and the voltage actually applied according to the scheme.

Since all functional connections, according to the Sampled Values and GOOSE communication services, are also described in the SCL language, they can be visualized in the software. For example, Fig. 7.23 shows the connections between the RPA devices of a 10 kV switchgear according to the GOOSE communication service. Each device is described, indicating its function (incoming line, sectional circuit breaker, outgoing line, etc.). The visualization shows characteristic communication patterns—for instance, it is evident that the sectional circuit breaker RPA device receives the most GOOSE messages, and the second-largest number of messages are received by the incoming sections of the 10 kV busbars.

The visualization of connections can also be detailed, showing the specific number of GOOSE messages sent and received by each device in the project, allowing the user to evaluate the correctness of the connections.

A key aspect for the verifier is not only the correctness of the communication (who is sending, who is receiving, and what type of message is being transmitted) but also the specific information being transmitted between devices. For the proper operation of protection and automation systems, certain signals (protection triggers, trips, etc.) are required. This information can be displayed and verified at a more detailed level, showing the objects and data attributes of logical nodes, i.e., their functions. The data objects and attributes in the IEC 61850 standard's semantics can be difficult to interpret, so software tools may use text descriptions of signals, defined at the specification stage, to facilitate user interpretation and support the verification process for ensuring the correctness of the connections. This also allows the identification of messages and signals that no devices in the project are subscribed to, i.e., messages erroneously included in the SCD file's electronic project documentation. Such messages and signals can lead to unnecessary loading of the local area network of the power facility.

The digital substation project is accompanied by a structural diagram of the PCS, reflecting the connection of secondary devices to Ethernet switches. The working

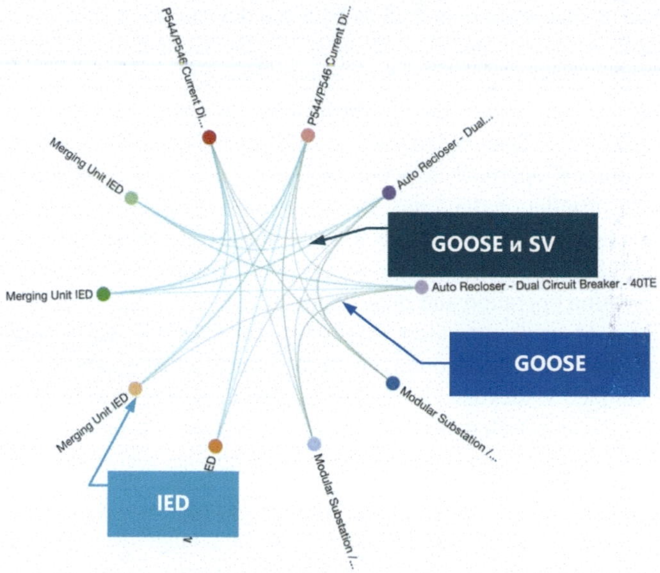

Fig. 7.23 Visualization of functional connections in the SCD file

documentation also specifies the switch configuration parameters, including the redirection of Sampled Values and GOOSE messages. The communication parameters for Sampled Values and GOOSE messages must align with the switch configurations to ensure that the required messages and signals are delivered to the intended recipients. These communication parameters are also described in the SCD file and can be visualized separately for each message. Analysing and adjusting these parameters (if necessary) will help avoid issues during commissioning.

Today, several companies develop and sell computer-aided design systems compliant with IEC 61850, which support the creation of SSD and SCD files. These include well-known manufacturers of secondary equipment, as well as companies not involved in developing intelligent electronic devices but specializing in software development. The most prominent examples include:

- HELINKS STS (Helinks LLC, Switzerland)
- ATLAN SCT (Pullnet, Spain)
- SCL Matrix (Gridsoftware, Canada)
- IET 600 (ABB, Switzerland), among others.

Control Questions and Tasks

1. Which chapters of the IEC 61850 standard regulate the design of digital substations using the SCL language? Explain the purpose of the Substation Configuration Description Language (SCL).
2. Name the five types of SCL files and provide a brief description of each.

3. Present the model and necessary explanations for information exchange during the design process according to IEC 61850.
4. What software tools are used to create SCL files? Provide examples of software that is used in practice.
5. What does the process of designing digital substations according to IEC 61850 look like?
6. Which SCL file describes the functional capabilities of intelligent electronic devices (IEDs)? What is the structure of this file?
7. What is the SCL file validation process? What tasks do specialists face when validating SCL files?

References

1. Humeres Flores P (2023) Planning and design for IEC 61850 implementation. IEC 61850 principles and applications to electric power systems. Springer International Publishing, pp 167–188
2. Steinhauser F (2020) Significance of high quality SCL data during the entire lifecycle of a protection, automation & control system. In: 15th international conference on developments in power system protection (DPSP 2020), pp 1–4
3. Aftab M, Suhail Hussain S, Ali I, Ustun T (2020) Novel SCL configuration method for modeling microgrids with IEC 61850. IEEE Syst J 14(2):2676–2683
4. Peng X, Liang Y, Luo Z, Pan F et al (2017) Analyses and comparisons of SCL files in substation configurator. In: 2016 3rd international conference on systems and informatics (ICSAI), pp 297–300
5. Schossig T, Naumann S (2014) Test automation utilizing SCL-files referring to IEC 61850. In: 12th IET international conference on developments in power system protection (DPSP 2014), pp 1–4
6. Himri Y, Muyeen S, Malik F, Himri S et al (2021) Review on applications of the standard series IEC 61850 in smart grid applications. Cyberphys Smart Cities Infrastruct Optim Oper Intell Decis Mak 197–253
7. Sun Y, Tang X, Li J, Jiang X (2013) Research on electrical power system with information modeling and SCL description of IED for transformer condition monitoring based on IEC61850. Adv Mater Res 676:177–180

Chapter 8
IEC 61850 Digital Communication Protocols

8.1 Data Communication Services and Protocols

One of the main ideas embedded in the IEC 61850 standard is its adaptability over time. The chapters of the standard progressively describe first the conceptual aspects of data communication within and between power facilities, then the so-called "abstract communication interface", and only in the final stage, the mapping of abstract models to data transmission protocols, which can evolve rapidly.

Thus, conceptual aspects and abstract models are independent of the data transmission technologies used (including communication protocols and transmission media), meaning they won't require revision as data transmission technologies progress.

8.1.1 Purpose of Abstract Communication Models and Services in IEC 61850

The abstract communication interface includes device models that standardize the concepts of "logical device", "logical node", "data object", and "data attribute", as defined in IEC 61850-7-4 and 7-3, as well as the description of data transmission services outlined in IEC 61850-7-2.

Section 7.2 describes more than 50 services related to client–server communications. In this case, an RPA device or connection controller acts as the server, and the SCADA system acts as the client. These services standardize the following procedures:

- Establishing communication between the client and server (Associate, Abort, Release);
- Reading the information model (GetServerDirectory, GetLogicalDeviceDirectory, GetLogicalNodeDirectory);

V. Y. Ushakov et al., *Digitalization of Electrical Power Engineering*, Power Systems,
https://doi.org/10.1007/978-3-031-95705-5_8

- Reading variable values (GetAllDataValues, GetDataValues, etc.);
- Transmitting variable values as reports (Report) periodically, sporadically, or on request;
- Tele control procedure (Control);
- Generating and transmitting event logs (Logs);
- Changing settings and settings groups (Settings Group Definition and Control);
- File transfer procedures, etc.

Reading services allow the client to retrieve the full information model from the device, recreating a tree of logical devices, logical nodes, data elements, and data attributes [1]. Thus, the client receives a complete semantic description of the data and its structure. Variable value reading services enable the retrieval of actual data attribute values, for example, through periodic polling. The report transmission service allows the setup of specific data transmission under certain conditions, such as changes related to one or more elements in the data set. The file transfer service can facilitate the transmission of COMTRADE format files.

For all client–server communications, speed is not as critical as the reliability and accuracy of the data exchange.

The first and most popular communication service in the "publisher-subscriber" category, shown in Fig. 8.1, is the Generic Object-Oriented Substation Event (GOOSE) communication service. The following characteristics apply to this communication service:

1. **A small amount of transmitted information**: status signals or analogue values.
2. **High-speed information transmission**. Since GOOSE communications are often used to transmit RPA signals and commands, the speed of this type of data exchange directly or indirectly impacts the response time for eliminating abnormal operating conditions. Therefore, the transmission of signals must occur with minimal delay.
3. **High Probability of Message Delivery** for critical functions such as sending a breaker trip command from protection relays (RPA) and exchanging signals between RPAs for distributed functions, guaranteed message delivery is essential. This must be ensured both during normal operation of the digital network and in the event of temporary network failures.
4. **Multicast Capability** for some distributed RPA functions, data needs to be transmitted from one device to multiple recipients simultaneously.
5. **Data Transmission Integrity Monitoring** the ability to monitor the status of the data transmission channel enhances the readiness coefficient during signal transmission, thereby increasing the reliability of functions dependent on these messages.

These requirements led to the development of the GOOSE message mechanism, which meets all specified criteria and will be discussed in more detail in Sect. 8.3.

The second communication service in the "publisher-subscriber" category is Sampled Values, which transmits instantaneous values from measurement transformers. The following requirements apply to this service:

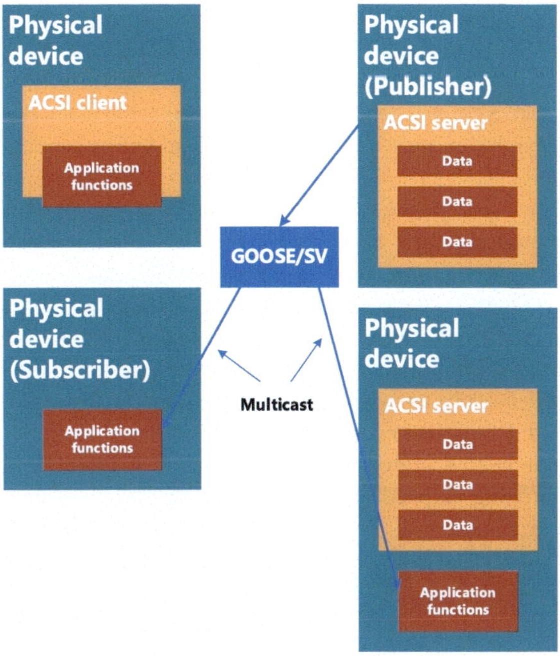

Fig. 8.1 Publisher-subscriber communication

1. **High-Frequency Data Transmission**. There is a need to transmit data at high frequency, leading to large volumes of information being transferred over the network.
2. **Minimal Data Transmission Delay**. Minimizing data transmission delay is essential, as this delay ultimately affects the response time of protection relay devices (RPA).
3. **Data Loss and Distortion Detection**. The ability to detect data loss or distortion during transmission is necessary to prevent unnecessary or false activations of RPAs and to provide signalling in the event of a communication channel or digital measurement device failure.
4. **Multicast Data Transmission**. One data stream may be required by various devices (e.g., voltage data from a busbar connected to a voltage transformer), so the ability for multicast data transmission must be ensured.

8.1.2 Communication Protocol Mapping

Now, we will discuss the process of mapping the aforementioned communication services to specific data transmission protocols, taking into account all the outlined requirements. In analogue signal transmission chains, the primary delay during transmission is caused by either the response time of the device's discrete input or signal filtering at the input. The time it takes for a signal to travel along a conductor is relatively negligible. Similarly, in digital data networks, the primary delay is caused not by signal transmission over the physical medium but by processing within the device.

Accordingly, and in consideration of the different requirements for various communication services, their mapping to specific communication protocols is performed. In data transmission network theory, services are segmented according to the levels of the Open Systems Interconnection (OSI) model, typically descending from the "Application" level, where data is represented in its final form, to the "Physical" level, where physical interaction between devices occurs. In the classical representation, the OSI model consists of seven levels: Physical, Data Link, Network, Transport, Session, Presentation, and Application, as shown in Fig. 8.2.

However, implemented protocols may not always incorporate all these levels, with some levels being skipped. This is because each conversion from one level to another introduces delays in forming a sequence of bits for transmission and in reverse conversion to retrieve the transmitted data. Therefore, to minimize delays, the number of conversions should be reduced to a minimum. For this reason, GOOSE and Sampled Values communication services, which are at the application level, are mapped directly to the Data Link level—Ethernet, bypassing other levels, in accordance with the demands placed upon them.

Client–server communications are mapped to the MMS protocol, as described by ISO/IEC 9506, and utilize the full TCP/IP stack, ensuring reliable connections between endpoints, albeit with lower responsiveness compared to GOOSE and Sampled Values communications.

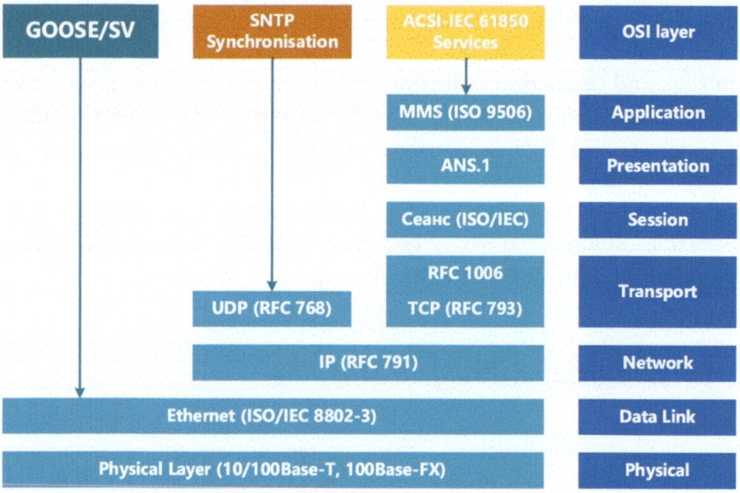

Fig. 8.2 Mapping to specific communication protocols

Based on the above, it is important to note that by default, data transmission using GOOSE and Sampled Values communication services is restricted to the local area network (LAN) of the power facility, meaning routing these messages beyond the local network is not possible. The rules and procedures for routing these messages are described in a separate chapter of the standard—IEC 61850-90-5. As for client–server communications using the MMS protocol, they are routable by default.

8.2 Application of Manufacturing Message Specification (MMS) Protocol

The IEC 61850 standard defines two communication models: (1) multicast communication and (2) client–server communication. Client–server communication is used for typical tasks in SCADA systems [2], such as remote control of switching devices, transmission of telecontrol and telemetry data, as well as logging and reading event logs and file transfers. This can also be used to transmit recorded waveforms. Client–server communication, according to IEC 61850-8-1, is implemented through the MMS (Manufacturing Message Specification) protocol [3]. MMS itself is not part of the IEC 61850 standard but is an external protocol used to implement abstract communication services defined by the standard.

8.2.1 Reports and Event Logs

The discussion of client–server communication in IEC 61850 begins with the transmission of reports and event logs. These two models cover the main areas in SCADA systems, including the transmission of telecontrol and telemetry information, as well as the logging and storage of event logs.

The reporting model, like the event log model, is based on data transmission or local storage in the device based on specific triggers. When a specific condition (trigger) occurs, data from a logical node is either transmitted to a remote client or stored locally in a log.

Unlike the periodic polling model used by many legacy data transmission protocols, which puts a constant load on the network and the device processors, the sporadic transmission model in IEC 61850 allows more efficient use of network bandwidth, minimizing device processor load, while nearly instantly transmitting data when events occur.

Key characteristics of the reporting and event log models are:

1. Sporadic reports act as a fast transmission mechanism, informing clients about the occurrence of specific events.
2. Event logs are stored on the server for later retrieval. When an event occurs, the server generates a report and logs it.

3. By using sporadic transmission or event-driven data, network load is minimized, avoiding continuous device polling. Thus, reports are only transmitted when necessary and can be precisely configured using various attributes, such as different criteria for report triggers.
4. The model provides the capability for data updates and system integrity checks at user-defined intervals or upon client request if necessary.

Key distinctive features of the report and event log models include:

- Immediate data transmission to clients upon a change in state, i.e., when a specific attribute's value or quality changes, as well as when analogue values exceed acceptable limits, etc.
- Event log storage ensures that reports are arranged in sequence, similar to event log entries. Even if there is a temporary communication loss, thanks to timestamps in reports, it is possible to reconstruct the sequence of events.
- Real-time reading of current values is available at any time through client requests to the server. Below, we provide a more detailed overview of the reporting model and its specifics.

The report transmission mechanism enables the transmission of packed data object values immediately after a specific event occurs, or after the expiration of a defined buffering time. The report transmission model in SCADA systems fully covers the field of telecontrol and telemetry data transmission.

The reporting model includes two types of reports—buffered and unbuffered—which account for the specific transmission of different types of data and allow for their use in various situations. Report transmission within the device is managed by the Report Control Block. It is important to note a key feature of these control blocks: each control block can only be associated with a single client. In other words, to send data to another system or device, a separate control block is required. Even if identical data is being sent to two systems, two separate instances of report control blocks must be created within the device, each transmitting data to its corresponding system.

As previously mentioned, there are two types of reports defined by the IEC 61850 standard: buffered and unbuffered. In the buffered reporting model, when a trigger activates the generation of a report, the report is transmitted immediately or stored in an internal buffer for later transmission if the client connection is unavailable at that time. As a result, no data values will be lost if communication fails when the report is being generated.

For unbuffered reports, buffering is not provided. When the trigger activates the generation of the report, it is sent immediately on a "best-effort" basis. However, if the connection is unavailable at that moment, the report may be lost. Based on this, we can conclude that the use of these models depends on the type of data being transmitted. When data integrity is critical, buffered reports should be used. If retaining all data is less crucial, unbuffered reports can be applied.

The abstract reporting model provides for two types of Report Control Blocks—Buffered Report Control Blocks (BRCB) and Unbuffered Report Control Blocks

(URCB). The model also includes a set of services that allow clients to read and configure the report control block parameters and directly manage the transmission of reports via the Report service.

The overall process for configuring report transmission is very similar to the configuration process for GOOSE or Sampled Values messages.

Firstly, a dataset of the data objects to be transmitted to the client system must be created. Then, the Report Control Block is configured, linking it to the previously created dataset, specifying the report identifier, and setting up the triggers for report transmission. The triggers for report transmission can include data changes (when data changes, a report is generated and sent); data quality changes (when a data object's Quality attribute changes, a report is generated and sent); and the Integrity trigger, which in combination with the report interval attribute, provides periodic report transmission at defined intervals. The General Interrogation (GI) trigger allows a report to be generated and transmitted upon a specific request from the client.

The reporting mechanism operates as follows. Unlike the GOOSE message multicast models, to activate report transmission, the report must not only be configured on the server but also activated on the client side. Therefore, the client first establishes a connection to the server and activates the Report Control Block. The client can also configure this control block if necessary. Once the report control block is activated by a specific client, it is reserved for that client, and no other client can connect to it. After this, reports are generated based on the configured triggers and transmitted from the server to the client that activated the report.

Configuration of Reports Using the SCL Language. To transmit a report, the device must be informed of the data to be included in that report. This is done by creating datasets. In Fig. 8.3, an excerpt from an SCL configuration file of a real device is presented, where some parameters are replaced. In this excerpt, two DATASET elements are of interest, named "meas_dataSet" and "TPU_DS", which show references to the data objects that will be transmitted in the report. Their creation is shown in Fig. 8.4.

It is important to note that for report transmission, data objects must be used, rather than specifying data attributes individually. For example, as shown in Fig. 8.4,

```
<IED name="W13G_Y001_001_AD_" type="ENMU-D" manufacturer="Energoservice" configVersion="1.0">
  ...
  <AccessPoint name="AP1">
    <Server>
      <Authentication />
      <LDevice inst="InternalLD">
        <LN0 lnType="InternalLD.LLN0" lnClass="LLN0" inst="">
          <DataSet name="meas_dataSet">
            <FCDA ldInst="InternalLD" prefix="Meas_" lnClass="MMXU" lnInst="1" doName="A.phsA" daName="cVal.mag.f" fc="MX" />
            <FCDA ldInst="InternalLD" prefix="Meas_" lnClass="MMXU" lnInst="1" doName="A.phsB" daName="cVal.mag.f" fc="MX" />
            ...
          </DataSet>
          <DataSet name="goose">
            <FCDA ldInst="InternalLD" prefix="IO_" lnClass="GGIO" lnInst="1" doName="Ind1" daName="stVal" fc="ST" />
            <FCDA ldInst="InternalLD" prefix="IO_" lnClass="GGIO" lnInst="1" doName="Ind1" daName="q" fc="ST" />
            ...
          </DataSet>
          <DataSet name="TPU_DS">
            <FCDA ldInst="InternalLD" prefix="IO_" lnClass="GGIO" lnInst="1" doName="Ind1" daName="stVal" fc="ST" />
            <FCDA ldInst="InternalLD" prefix="IO_" lnClass="GGIO" lnInst="1" doName="Ind2" daName="stVal" fc="ST" />
            ...
          </DataSet>
```

Fig. 8.3 Datasets

```
<ReportControl name="MEAS_TPU' datSet="meas_dataSet" intgPd="1000" rptID="1_W13G_AD_Y001_05_PM_U" confRev="1">
  <TrgOps period="true" />
  <OptFields seqNum="true" timeStamp="true" reasonCode="true" configRef="true" />
  <RptEnabled max="1" />
</ReportControl>
<ReportControl name="TPU" datSet="TPU_DS" intgPd="10000" rptID="2_W13G_AD_Y001_05_AD_B" confRev="1">
  <TrgOps dchg="true" qchg="true" period="true" />
  <OptFields seqNum="true" timeStamp="true" reasonCode="true" configRef="true" />
  <RptEnabled max="1" />
</ReportControl>
```

Fig. 8.4 Report control block

in the "meas_dataSet" dataset, there is a reference to the logical device "ldInst"; the prefix of the logical node, such as "Meas_"; the logical node class "MMXU"; the instance number "lnInst = 1"; and "doName" with the name "A.phsA". However, no specific data attribute to be transmitted is indicated—the server automatically selects the necessary data attributes based on the Functional Constraint (FC) attribute (this is another attribute of the data object reference included in the dataset). In this case, "fc = MX" means that all attributes corresponding to "Functional Constraint = MX" in the data object should be included in the dataset and transmitted to the client.

This mechanism is crucial for several reasons: in contrast to GOOSE or Sampled Values messages, only the data that triggered the event is transmitted in a report. Thus, if we were to include data attributes individually in the dataset, only those attributes that triggered the event would be transmitted. Therefore, we wouldn't be able to, for example, simultaneously retrieve both the data value and the quality flag. However, when transmitting an entire object, the trigger applies to the entire object, allowing us to receive both the data value, the quality flag, and the timestamp for these data in a single report.

Further, the configuration of the Report Control Block will be discussed. In the SCL file, the Report Control elements are provided for this purpose. The Report Control element includes attributes such as: its name, a reference to the dataset (the dataset's Name attribute, as shown in Fig. 8.1), the report identifier rptID, and the configuration version of the report, which must be updated every time the dataset it references changes.

Next, within the appropriate ReportControl element, the TrgOps element is defined, with attributes specifying the triggers to be used for transmitting the report. In this case, we see that all three triggers are utilized: data change, i.e., a change in data; qchg, a change in quality; and periodic transmission at defined time intervals.

Within the OptFields element (optional fields), additional fields that should be included in the report, aside from the data itself, can be specified. Such fields may include the ReportID—the report identifier—and other parameters that may be added as per the standard.

The RPTenabled element with the attribute "max = 1" indicates that one instance of this report will be activated on the server. While there is only one report control block in the SCL configuration, one instance will be activated in the device model. Thus, this report control block can send data to five clients.

Now, the mapping of the Report service to a specific communication protocol, in accordance with Sect. 8.1 of the IEC 61850 standard, should be mentioned. The protocol used is MMS. Mapping the report service to the MMS protocol is highly technical and of more interest to developers than to operators or experts, so we will summarize this briefly.

The abstract report generation service is mapped to the Information Report service of the MMS protocol. The Information Report service in MMS is identical to the Read Response service (which responds to data read requests from the client), but it is generated spontaneously by the client, i.e., without a server request. The Information Report transmits a Named Variable List—a standard MMS object (which can be translated as a "list of named variables"). The response also contains the Variable Access Specification attribute, which indicates that it is a report (RPT), and the so-called Access Results, which contain the data of the transmitted report.

In conclusion, the recommendations for applying these different report types will be presented. It is important to note that measurements should generally be transmitted via unbuffered reports. This is because for measurements, the value lies in real-time data rather than buffered data. Furthermore, frequent changes in measurement values may quickly overflow the buffer in case of a communication failure. Discrete signals, including protection activation signals, circuit breaker states, etc., should be transmitted via buffered reports, ensuring that the data is preserved even in the event of a communication failure. The event log on the client side must always have accurate timestamps, and therefore, discrete signals should be transmitted via buffered reports.

To update parameter values transmitted via unbuffered reports after a connection is restored, the GI (General Interrogation) trigger or a client-initiated general poll should be used to retrieve the current values of all data being transmitted in the report. Additionally, for analogue values, reports can be configured to be sent periodically using the Period attribute, ensuring that analogue values are updated on the client side at regular intervals even if they have not changed.

The final recommendation is that data objects should be used in the datasets for report transmission, rather than specifying data attributes directly. Otherwise, the report transmission mechanism may not function correctly, and valuable data may not be received in certain conditions.

Next, we will examine the application area and general characteristics of event logs used within the IEC 61850 standard. Event logs, like those in any SCADA or protection and control systems, are used to store event information in the internal memory of a device for subsequent reading. The event log model is essentially identical to the buffered report, except that the report is not immediately sent to the client but is saved in the device's internal memory for later transmission. Unlike the reporting model, event logs can be accessed by multiple clients. Moreover, this model provides more flexible reading options for specific events, such as by date or date range.

The diagram in Fig. 8.5 illustrates the operation of the abstract event log model. The key elements in this diagram are: LOG—the event log itself; LCB "A" and LCB "B", which are the event log control blocks; and the logical node GLOG, which can also be used to store entries in the event logs. These elements are situated within two logical devices—LD A and LD B—inside the physical device (the server).

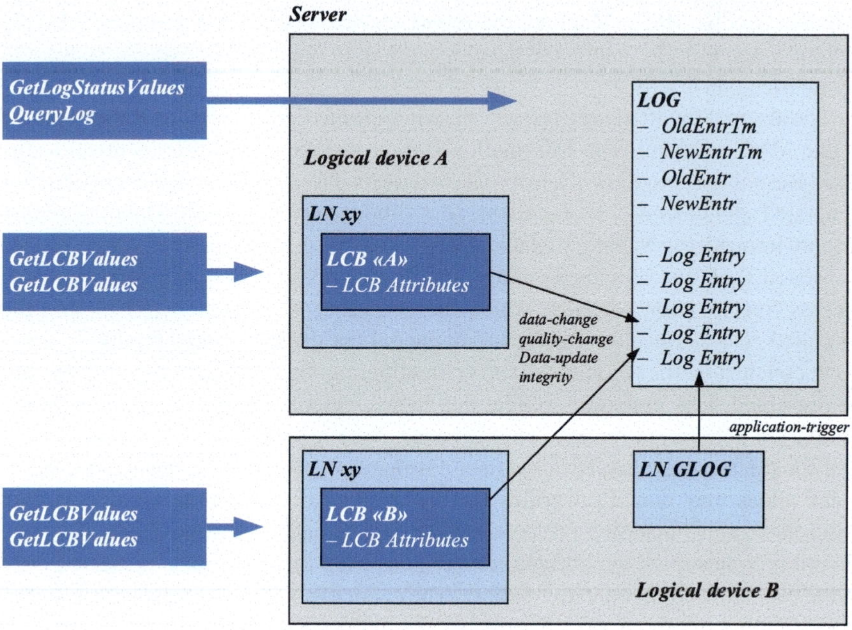

Fig. 8.5 An abstract model of the event log

Each event log control block (similar to the report service and report control block) refers to a specific dataset, with triggers forming entries within the log. Triggers include data changes, quality changes, data updates, and periodic logging. Additionally, for storing data related to the moment of trigger occurrence, but which does not directly cause the trigger, the GLOG logical node can be configured to log different data, triggered by changes in other data.

The LOG event log contains entries about these events, as well as attributes that can be read, such as the time of the most recent entry (NewEntrTm), the time of the earliest entry (OldEntrTime), and the identifiers of the oldest (OldEntr) and most recent (NewEntr) entries in the log.

This model also includes services—some are directly related to the event log, while others pertain to the control blocks that manage log data storage. The event log services (GetLogStatusValues and QueryLog functions) allow reading the current attributes of the log, such as the times of the oldest and most recent entries and their identifiers. The LCB services (GetLCBValues and SetLCBValues) allow reading the current parameters of the control block and writing new parameters to reconfigure the block.

Further, the event log services will be considered in more detail. The first service is QueryLogByTime, which retrieves log entries over a specified time interval. The parameters for this request include the start and end times of the interval. The next service, QueryLogAfter, retrieves log entries starting from a specific point in time or after a particular entry identified by its EntryID. This is useful, for example, when a client has previously requested the event log and retained the identifier of the last entry received; the client can then request all subsequent entries. The final service, GetLogStatusValues, retrieves the current status of the event log control block.

Event logs are used to store event information even when communication is unavailable or not required. Additionally, event logs can be used when communication channels do not have sufficient bandwidth to transmit reports as frequently as the device generates them. Lastly, event log services can be useful when the priority is not real-time signalling but the preservation of information about the sequence of events.

8.2.2 Operational Control

Another crucial aspect of communication protocols mapped to the MMS protocol is operational control [4]. The operational control model enables the client to alter the states of internal and external processes within the server. This applies not only to the operational control of switching devices but also to any other processes, such as resetting LED indicators or issuing other control commands. The control model is applicable only to data objects derived from the general class of controllable data—Command Controllable Data Class—and where the network attribute ctlModel is not set to "status-only" This means that if a data object is theoretically controllable but has its ctlModel set to "status-only", it cannot be controlled. Data objects that can be managed are referred to as Control Objects.

Figure 8.6 presents an abstract control model, with the central area occupied by the control object itself, or, in the context of the IEC 61850 standard, a data element (or the data itself according to the second edition of the standard). The control object is not a physical entity but a logical data object. This control object has several attributes: ctlVal—the control value; operTm—the operation time for time-operated services; Origin—an attribute that indicates the source of the command; ctlNum—the control command number; and several status values that are not controllable, including stVal—the current value of the controlled variable, t—the timestamp, and q—the quality flag.

The control model operates as follows: Operator A (the client system) issues an Operate command with the control value ctlVal Open (i.e., "open") along with a timestamp (t) and a delay for the command execution, identifying themselves as "Operator A". In response, the controlled object acknowledges the receipt of this command. Once the command is received and the control object starts execution, the final control command, shown on the right side, is initiated. This final automation process lies outside the scope of the standard since each manufacturer defines it independently. As a result, the command may activate some physical device or modify logical values within the system if no physical control is involved. After the command is executed, the control object generates a report indicating the outcome of the operation through the corresponding attributes.

The standard defines four control models:

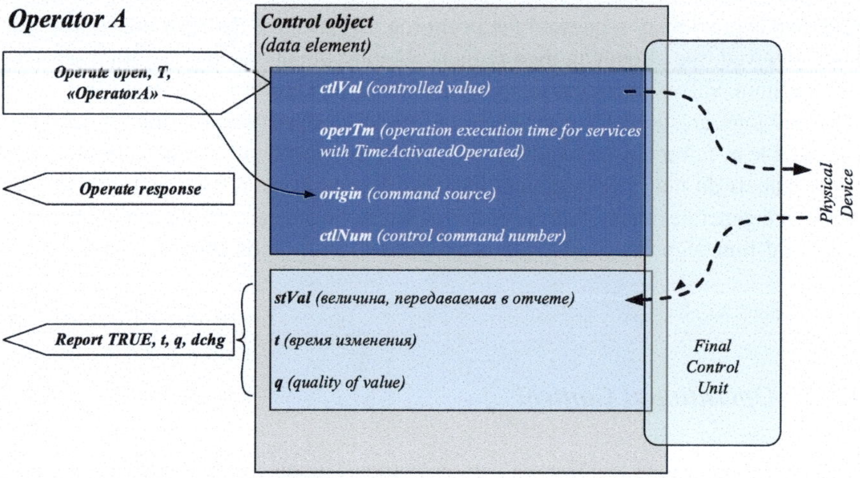

Fig. 8.6 Abstract control model

(1) Direct control without feedback
(2) Direct control with feedback
(3) Select-before-operate control without feedback
(4) Select-before-operate control with feedback.

Each of these models has its specific characteristics and corresponding application areas. Direct control without feedback is used for the simplest operations, which do not involve external equipment state changes. In contrast, select-before-operate control with feedback is applied for more complex and critical operations, where monitoring the position of primary or physical equipment controlled by the object is required.

Let us first look at the simplest model—*direct control with normal security* (Fig. 8.7a). This model is used to control elements or variables, typically local data objects within the device itself, where a certain attribute value is set without impacting external equipment. Since feedback is not provided, the control is performed independently of other objects. Examples of such controls include resetting LED indicators, turning on heaters, or starting ventilation systems. The client issues a command directly to the server, which confirms receipt and reports the outcome.

Fig. 8.7 Control models: **a** direct control with normal security, and **b** select before operate control with normal security

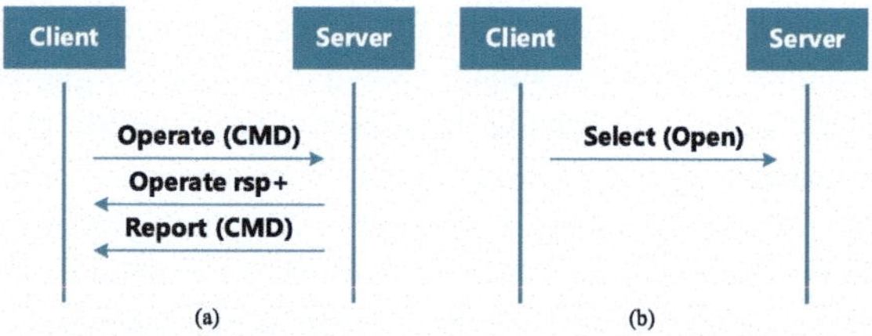

The next control type is *select before operate control with normal security* (Fig. 8.7b). This model includes an additional step for selecting the control object. In this case, issuing a control command is impossible without first selecting the command. The client sends a Select request to the server, and after receiving confirmation, it can send an Operate request, which executes the command. Upon receiving the Select request, the server grants exclusive control of the data object to the client, preventing other clients from issuing the same command. Any further Select or Operate requests from other clients will be rejected.

The next two control models are direct control and select-before-operate control with feedback.

The control model with enhanced security implies that during the execution of the control command, feedback from the mechanism or device controlled by the data object is monitored. In feedback control models, each control cycle must conclude with a notification indicating the completion of the control cycle, referred to as Command Termination. This model is applied in cases where the control procedure has significant consequences for external devices or equipment. After the command is issued, the state of the equipment must be verified to ensure that it has changed to the desired new position. For example, in the case of circuit breaker control, it is important that its position changes to the new (desired) state after the command is executed.

Figure 8.8 shows the cycle of *direct control* and *select before operate control with enhanced security*, which is the most complex. When using this model, the client first sends a Select request to the server and receives confirmation that the command can be executed. After that, the client sends an Operate request to the server. The server first acknowledges receipt of the command and then confirms that it has begun executing it. The command is then sent to the equipment, the actuator is activated, and its state starts to change. An intermediate state is reached. Once the actuator has completed its task, the state changes to the "open" position (for example), and the server sends a report indicating that the command has been executed. After this, the server releases the equipment and sends a Command Termination signal, indicating that the command has been completed.

It is worth noting the mapping of control model services to a specific communication protocol, such as the MMS protocol. To implement this procedure, two main services are used: Write and InformationReport. The selection, i.e., SelectWithValue, and the control command, Operate, are mapped to the Write service, meaning that these commands effectively function as writing a specific value to a specific variable. The server's response indicating the inability to execute a given command (Response-) with the reason for the failure is mapped to the InformationReport service of the MMS protocol.

The control models applicable to specific data objects are typically determined by the device manufacturer and are not configurable by the user. Usually, the user cannot independently change the control model for a particular data object. However, it is useful to understand which objects are used and where, to monitor what the manufacturer has implemented in certain cases.

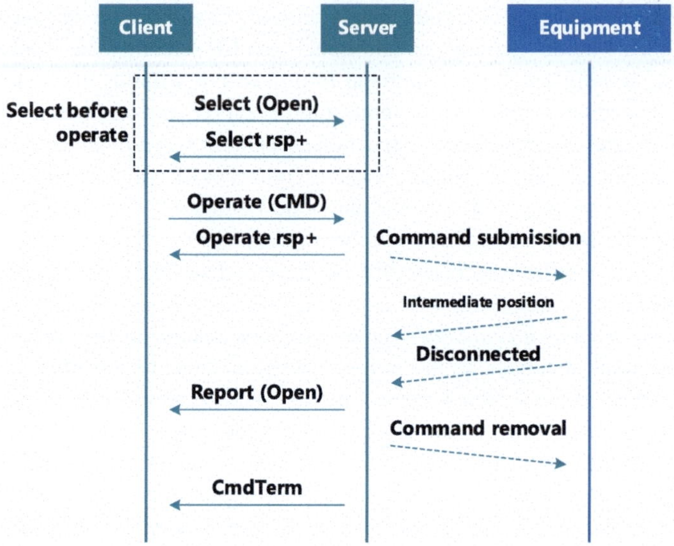

Fig. 8.8 Control models: **a** direct control with normal security, and **b** select before operate control with normal security

For simple operations (e.g., resetting LED indicators, turning on lighting, heating, or cooling—where, on the one hand, the operation does not require control from multiple clients, and on the other, the execution of the command is not monitored, i.e., no feedback is provided), the simplest model, direct control without feedback, is typically applied. For critical operations involving electrical switching, where it is important to monitor the execution of the operation, i.e., to provide feedback (e.g., for the remote operation of circuit breakers, disconnectors, grounding switches, tap changers, etc.), the select-before-operate control model with feedback is usually applied.

8.2.3 Changing Setpoints and Setpoint Groups

It should be noted that each instance of a data object can hold only one value at any given moment. This means that, in general, we can assign only one value to the settings of functions (via the data objects of the IEC 61850 information model). In practical applications, particularly in relay protection, it is often necessary to store different parameters for different settings of the same data object and to change them for a group of objects simultaneously. To address this need, the setting group control model is provided, allowing for such functionality. Furthermore, it offers the ability to modify individual settings dynamically from the client side.

Setting group management is performed via the Setting Group Control Block, which can only exist once for each logical device. Like other control blocks, it is located in the LLN0 logical node. The number of setting groups supported by a device is determined by the manufacturer and specified in the accompanying device documentation, particularly in the PIXIT document. To enable the client to determine which settings are active, a separate Functional Constraint corresponding to SG is provided. If the data model shows parameter values with the Functional Constraint SG, it means that these values belong to the active setting group. Additionally, another Functional Constraint is provided for set points that are in the editing buffer, denoted as SE.

It is important to note that even if settings are not described in the IEC 61850 data model (e.g., protection settings), the manufacturer can still implement a setting group control block and manage the setting groups, with changes being made using dedicated engineering software. The traditional setting group can be switched using the Setting Group Control Block.

In the abstract model for managing setting groups, the central component is the Setting Group Control Block (SG Control Block), as shown in Fig. 8.9. In the example shown, there are three setting groups, and switching between the active groups is done using the SelectActiveSG service. The Setting Group Control Block is depicted as a switch for selecting the active group using the SelectActiveSG service. This service sets the active group, and the values from this active group are applied to all functions within the logical device.

The next service is SelectEditSG, located at the bottom right, which is used to transfer one of the selected setting groups into the editing buffer. After the SelectEditSG request is sent from the client to the server, specifying the corresponding setting

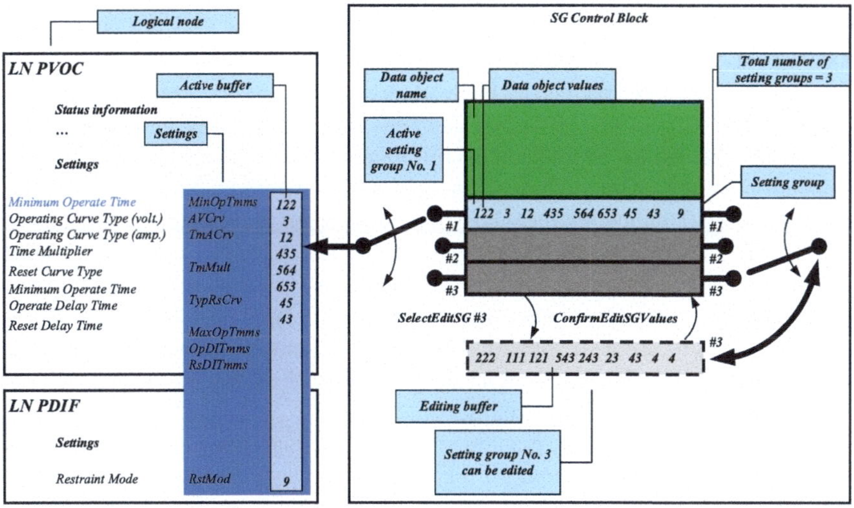

Fig. 8.9 Setting group control block

group, the server moves the settings from that group to the editing buffer, where they can be modified using the SetEditSGValue and GetEditSGValue services to read or write the respective values.

Once the necessary settings have been modified, the ConfirmEditSGValues request is sent from the client. As shown in Fig. 8.9 (arrow pointing up), the server transfers the edited set points from the buffer back to the setting group, thus completing the editing process. If the active setting group is edited, the settings are updated all at once only when the ConfirmEditSGValues request is sent. As mentioned earlier, a single setting group control block applies to all logical nodes within one logical device, meaning that all logical nodes of a device are edited via the same setting group control block (as shown on the left in Fig. 8.9).

The Setting Group Control Block includes the following attributes:

1. NumOfSG—the number of setting groups in the corresponding node;
2. ActSG—the number of the active setting group;
3. EditSG—denotes the group number that is copied into the editing buffer;
4. CnfEdit—edit confirmation; its value is set when it is necessary to confirm changes to the setting groups;
5. LActTime—the time of the last setting group change;
6. ResvTms—the time for which exclusive control over the Setting Group Control Block is granted to a specific client.

Settings for the group being edited can be written and read using the GetEditSG-Value and SetEditSGValue services, respectively. Changes to the setting group will only take effect if the ConfirmEditSGValues command is issued. If the command is not issued within the ResvTms period, all the changes made will be discarded, and the device will continue operating with the previously selected settings.

In addition, the mapping of the setting group control model services to a specific communication protocol, such as MMS, in accordance with Sect. 8.1 of the IEC 61850 standard should be discussed. Firstly, all services in the Get format (i.e., retrieving any data) are assigned to the Read service of the MMS protocol. Similarly, all Set services (i.e., setting or writing data) are assigned to the Write service.

The PIXIT document should contain information about the maximum number of setting groups supported by the device in question. The initial value of EditSG must always be set to 0, indicating that no setting group is currently in the editing buffer. Finally, only one class, one Setting Group Control Block, can exist within a logical device.

The first and most important recommendation for applying the setting group control model is to use it (if supported by the manufacturer) instead of proprietary functions in the manufacturer's engineering software or custom-programmed logic for changing settings or setting groups.

Another key point is that the greatest efficiency in using setting groups is achieved when all settings, including those for relay protection, are described in the SCL file and later reflected in the MMS tree, which can be read by the SCADA system. In other words, combining the description of settings within the device's information model, using IEC 61850 communication services, and employing the setting group control model allows for the most efficient use of this service.

It is important to note that the comprehensive use of IEC 61850 communication services enables the transition to universal software-hardware solutions for managing and editing settings, independent of specific manufacturers of protection and automation devices (PACs) and SCADA systems. Settings, including their values and groups, can be changed without relying on proprietary software tools from individual manufacturers, using a universal method applicable to all devices.

8.3 Generic Object Oriented Substation Event (GOOSE) Protocol

The GOOSE protocol (Generic Object Oriented Substation Event) is one of the most well-known protocols defined in IEC 61850 in Sect. 8.1. In general terms, its purpose is to facilitate the exchange of information between protection and control devices (PACs) in a digital format [5]. GOOSE protocol provides a fast event transmission mechanism and is used for breaker trips, triggering events, recording fault events, and more.

GOOSE messages are used to replace traditional copper cabling used for transmitting discrete signals between devices. Traditional communication protocols commonly used in the energy sector, such as Modbus and IEC 60,870, do not provide the necessary speed or, more precisely, the low latency required by protection and automation systems [6]. One of requirements for GOOSE message transmission is minimal latency so that this communication service can be applied to protection functions and sending breaker trip commands. To fully understand GOOSE, it is important to grasp the basic principles underlying this model, as defined by the standard.

First, it is essential to refer to Sect. 7.2 of the standard, which introduces the concept of a Generic Substation Event (GSE). This model is divided into two submodels: GSSE and GOOSE, as illustrated in Fig. 8.10.

The GSE model, described in Sect. 7.2 of IEC 61850, provides a means for fast and reliable transmission of values across the entire secondary system. Essentially, GSE is a model for disseminating information about an event occurring in a substation, station, or any other energy facility in a broadcast or multicast mode, meaning the

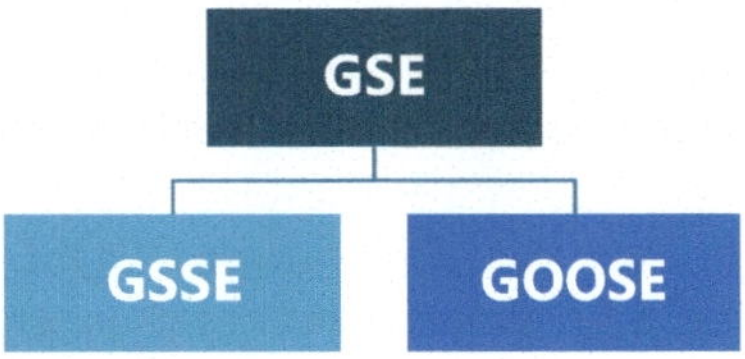

Fig. 8.10 General event at the substation

information is transmitted from one device to multiple devices. A specific case of this is sending information from one device to only one other device.

The Generic Substation State Event (GSSE) is a sub-model that handles general state information at a substation. This model provides the ability to transmit state changes through the transmission of bit pairs, enabling the transfer of discrete signals like logical zero or one. This model was the first introduced and included in the original edition of IEC 61850, and it was initially defined in the UCA 2.0 document, a predecessor of IEC 61850. However, practical implementation of the GSSE model in devices has been virtually non-existent. As of the second edition of the standard, GSSE is considered obsolete, moved to an appendix, and is not recommended for use.

The second model within the larger GSE model is GOOSE, or Generic Object Oriented Substation Event. Unlike GSSE, GOOSE is an object-oriented event, meaning it allows for the transmission of more than just discrete signals, such as bits or bit pairs. GOOSE can carry any data described by the IEC 61850 information model, including integers, floating-point values, and even logical values like "zero" and "one", In short, GOOSE is not limited to transmitting logical (Boolean) data but can send a wide variety of data types.

8.3.1 Abstract Model, GOOSE Message Transmission Block, and Communication Services

Before diving into the abstract model of GOOSE, it's important to remember that abstract models in IEC 61850 are designed to ensure independence between various chapters of the standard and technological advancements in data transmission. In this way, the abstract GOOSE model describes the message transmission mechanism without being tied to specific data transmission technologies. The abstract GOOSE model is provided in Sect. 7.2 of the standard.

The abstract GOOSE model only outlines the principles of message transmission, without focusing on how these principles are implemented through specific communication protocols and technologies [7]. The model contains the essential principles and required attributes for data transmission. However, each specific protocol implementation, or protocol mapping, may add additional attributes necessary for that particular implementation.

The mapping of GOOSE messages to a specific protocol is covered in Sect. 8.1 of the standard, where GOOSE is mapped to the data link layer using the Ethernet protocol. This approach is essential for separating abstract principles from specific communication protocols, ensuring that the principles do not change as data transmission speeds evolve, which is progressing rapidly.

After that, the abstract GOOSE message transmission model is discussed. In Fig. 8.11, the publisher of the message is shown on the right, while the subscriber is on the left. The publisher has data in its information model that needs to be transmitted

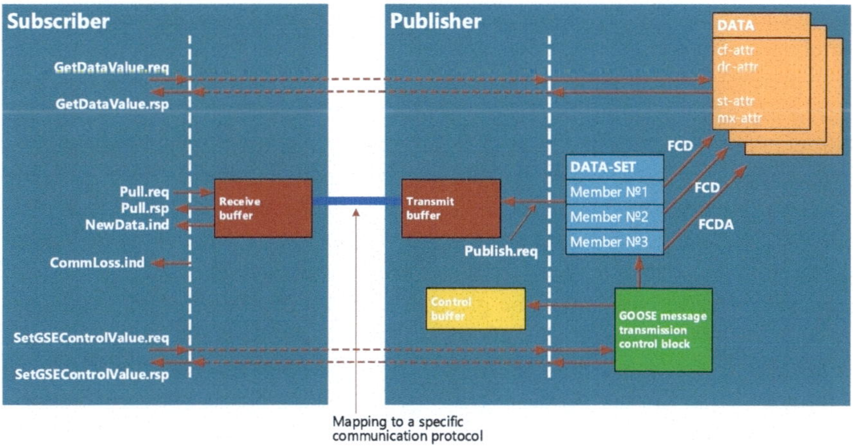

Fig. 8.11 Abstract model of GOOSE message transmission

via a GOOSE message. These data are represented in the upper right corner as DATA. To send the message, the data are grouped into a data set (DATA-SET), shown below and to the left of the DATA. We will discuss data set grouping later.

At the bottom of the publisher is the GOOSE Control Block, responsible for managing GOOSE message transmission. Then, there is the transmission buffer, where the data are mapped to a specific communication protocol, such as the protocol used to transmit data from the publisher to the subscriber. Finally, the subscriber has a receive buffer and an application communication interface, which each manufacturer implements uniquely.

Thus, the GOOSE message transmission model consists of several core elements, which, according to the standard, must be consistent across manufacturers. These elements include:

- DATA: The description of the information model, representing the data to be included in the message.
- DATA-SET: A collection of data grouped for transmission within the message.
- GOOSE Control Block: Manages GOOSE message transmission.

The GOOSE Control Block is used to configure GOOSE transmission, requiring a set of parameters, one of which is the data set. A data set is simply a collection of references to data within the device's information model. All communication services in the standard that involve transmitting data externally, whether to other devices or a client system like a SCADA system, operate using the concept of a data set.

Figure 8.12 shows a data set that includes the positions of switching devices for one connection. Each switching device, such as those inside a feeder controller responsible for managing that connection, is associated with a logical node. In this case, the logical nodes XSWI represent switching devices like isolators or grounding switches.

To transmit the positions of these devices using GOOSE messages, a data set must be created. In this example, the data set representing switch positions is shown. The data set is always designated as DATA-SET (a universal term found in all configurators and SCL files). The data set contains references, and in the GOOSE control block, a reference to the data set is specified. This informs the device that when transmitting messages, it should reference the specified data set, identify the data to be transmitted, and then transmit the data linked in that data set.

Further consideration should be given to a critical feature of GOOSE message transmission—the high reliability of data transmission while maintaining maximum speed. One of the ways to ensure reliable GOOSE message transmission is the continuous retransmission of messages in both normal mode and post-event mode. Continuous retransmission is a mechanism for constantly diagnosing the communication channel, such that after each specified time interval T0, the device transmitting

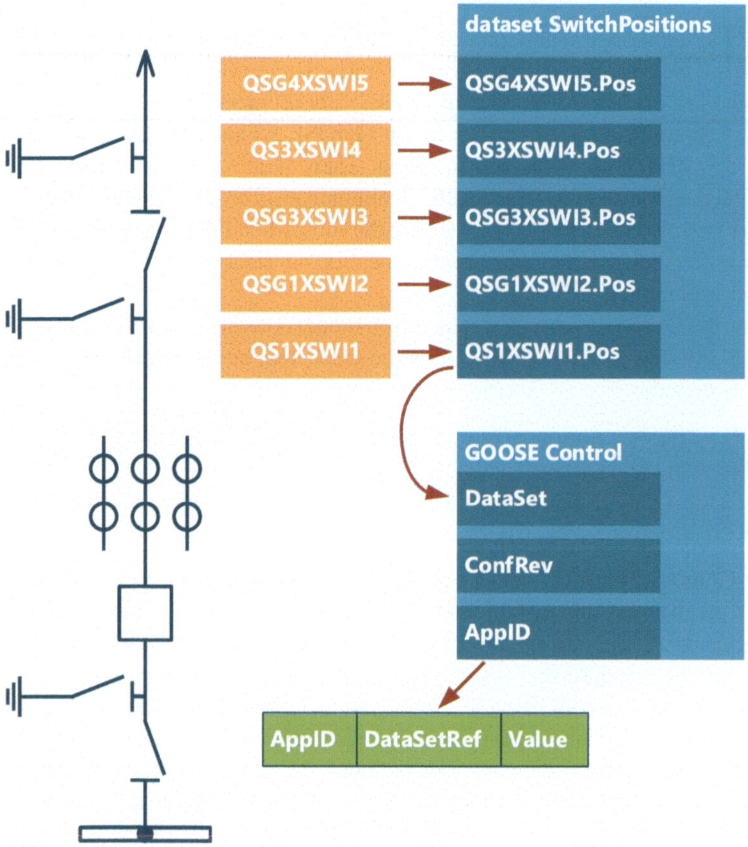

Fig. 8.12 GOOSE control block

GOOSE messages must send the next message, even if no changes have occurred. This means that messages are sent with the same values as before, except for the service counters (in this case, the sqNum counter, which increments by one with each transmission). This allows receiving devices to confirm that the communication channel is functioning properly without any issues by receiving these messages at interval T0.

At the moment when an event occurs, such as a protection trip, protection start, or switching of a circuit breaker, a message is sent immediately. It does not wait for the next cycle; instead, it is sent instantly. The sqNum counter is reset, and the stNum counter (which tracks the number of status changes within the messages) is updated accordingly, as shown in Fig. 8.13.

This model also includes an interesting mechanism that ensures quick delivery of information to receiving devices even in the event of a brief communication failure. This is achieved by significantly reducing the message retransmission interval after an event. Typically, the minimum interval T1 is around 3–4 ms for most manufacturers, with some even offering lower values. This allows a second or third message to be delivered within 10 ms if the first message after an event is "lost" for some reason. On one hand, this ensures constant channel diagnostics, while on the other, it provides reliable message delivery to receivers even in the event of brief communication disruptions.

It is important to emphasize the significance of the continuous message retransmission mechanism. Messages are continuously sent at equal intervals, even when no events occur, enabling receiving devices to continuously self-diagnose the communication network. In the GOOSE mechanism, there is no acknowledgment from the

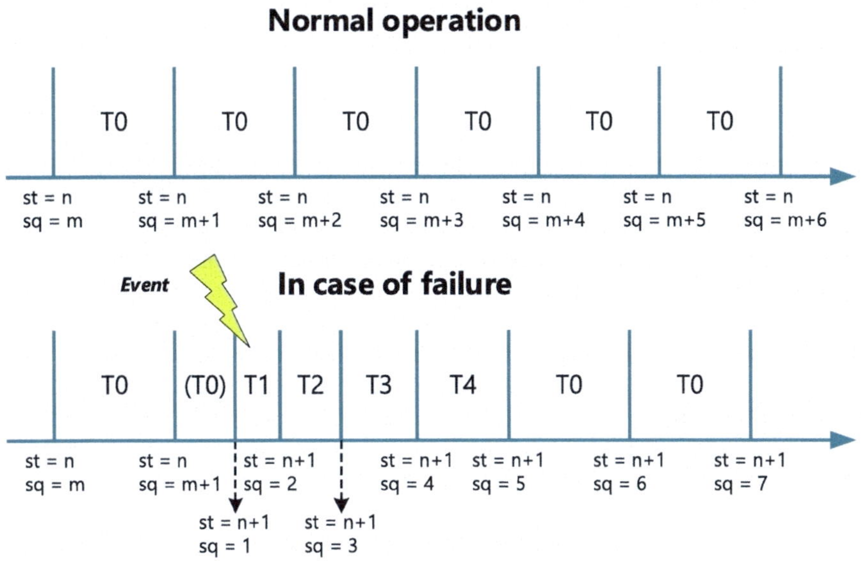

Fig. 8.13 Ensuring data transmission reliability for GOOSE messages

receiver confirming that the sender's message has been received. In this context, such a mechanism is fundamentally impossible because there are often many receivers, and it would be infeasible to track delivery to all of them. Instead, receiving devices are responsible for diagnosing the communication channel, and if a message is not received for some reason, they signal this. This continuous diagnostic capability significantly enhances the reliability of data transmission and system readiness overall.

Another key point to note is the retransmission period in normal mode and post-event mode. This period is a configurable parameter consisting of MaxTime and MinTime. The standard does not regulate the values of these parameters. The MinTime is typically limited by the manufacturer and ranges from 3 to 5 ms for most manufacturers, with some offering values as low as 1 ms. The value of MaxTime can either be defined by the standard profile used by the organization or determined by the specific project. MaxTime should be chosen based on a calculation of system reliability (availability) and the consequences of not detecting a failure within the MaxTime period. Typical project values for MaxTime are around 5–10 s, though in some cases, it can be as low as 1 s. It is important not to set MaxTime too low, as this would unnecessarily increase network load without providing significant benefit.

It is also crucial to remember that while GOOSE includes built-in diagnostic capabilities, this does not automatically guarantee system reliability. It is important for the project to not only be aware of this diagnostic capability but to use it effectively. This can be done, for example, by implementing specific measures to monitor communications and quickly alert personnel to any issues in the communication network. In essence, the system's reliability, or more precisely, its availability, is determined by three factors:

- The reliability or failure-free operation of the equipment used.
- The speed at which failures are detected.
- The speed at which failures are resolved.

The GOOSE protocol cannot influence points 1 and 3; equipment reliability falls outside the scope of the IEC 61850 standard, and the GOOSE mechanism cannot affect the speed at which faults are repaired. However, GOOSE can control how quickly failures are detected by using the mechanism described above.

Separate attention should be paid to the key attributes of the GOOSE Control Block, as shown in Fig. 8.14. The first attribute is the GOOSE Control Block Name (GoCBName), which is the name of the GOOSE control block. Typically, the GoCBName parameters are fixed and predefined by the device manufacturer, meaning they cannot be modified in most cases. These parameters belong to the device model and are not configurable.

The next parameter is GoCBRef, a reference to the control block within the device's information model. This reference is determined by both the name of the control block and its location in the logical node, which indicates the block's relative position within the device. This reference is also non-configurable and is determined by a set of parameters.

№	Parameter	Description
1	GoCBName	GOOSE control name, identifies a GoCB within the scope of a GoCBRef (GOOSE control reference)
2	GoCBRef	Describes the GOOSE control block that controls the message
3	GoID	User-assigned identification for the message
4	DatSet	Reference of the data set whose values of members shall be transmitted
5	ConfRev	Configuration revision indicating the number of times the configuration of the data set referenced by DatSet has been changed. The counter is incremented every time the configuration changes
6	DstAddress	Attribute that specifies the destination address

Fig. 8.14 Key attributes of the GOOSE control block

The third parameter is GoID, a crucial parameter that serves as the identifier for GOOSE messages. GoID allows receiving devices to identify the source device and understand the purpose and content of the message. Therefore, it is important that in projects using GOOSE messages, GoID is assigned based on clear design rules indicating the originating device and the intended use of the message.

The next parameter is DatSet, which in GOOSE control blocks refers to the data set to be transmitted. This parameter represents a reference to the data set in the device's information model, specifying which data will be transmitted in the GOOSE message.

Following this is the ConfRev parameter, representing the configuration version. This is another critical parameter, used in GOOSE message transmission (and similarly in Sampled Values transmission) to ensure that receiving devices are using the same configuration version as the sender. When the system is configured, each device is loaded with its configuration, including the data and attributes it will send. The receiving devices are also loaded with a configuration that specifies which messages and attributes to receive. Therefore, it is essential that all devices use the same configuration version, controlled by the ConfRev attribute. If the sender transmits a message with configuration version 5, but a receiver still uses version 4 due to a configuration error, the receiver should signal a parameterization error. How this is implemented varies by manufacturer.

Finally, the Destination Address attribute defines the destination address. Even though GOOSE uses multicast, a destination address must still be specified. This is known as a group address, and its primary purpose is to filter and restrict the spread of messages across the network, ensuring they do not reach unnecessary segments.

8.3.2 Configuration of GOOSE Message Transmission in SCL Language

As mentioned earlier, the SCL language is used for configuration in accordance with IEC 61850, and the transmission of GOOSE messages is no exception to this rule.

```
<GSEControl desc="Predefined GOOSE ControlBlock" name="gcbBasicGse" datSet="gseDS"
confRev="10001" appID="1_T1H_PA_S002_A1_PS">
  ...
</GSEControl>
```

Fig. 8.15 Description of GSEControl element

The procedure for configuring GOOSE communications in SCL follows a standard sequence: firstly, a data set is created, then the GOOSE Control Block is configured with its communication parameters. After the GOOSE message transmission is configured, the receiving device must be set up to subscribe to this GOOSE message [8]. In terms of GOOSE message transmission, it is important to note that the configuration of the control unit itself (this is the GSE Control element within the IED element) is done independently of its communication parameters, which are located in the GSE block within the Communication element. This separation ensures that the functional configuration of the GOOSE Control Block can be handled by protection engineers, while communication parameters like the MAC address VLAN, etc., can be managed by specialized communication staff. This division of responsibilities allows for efficient management of both the protection and communication functions.

Consider the GSE Control element within the IED element in Fig. 8.15:

- GSE Control—This element corresponds to the control block responsible for the transmission of GOOSE messages and includes the following attributes in the SCL file:
- Name—the name of the control block.
- Datset—a reference to the data set that will be transmitted in the GOOSE message.
- ConfRev—the configuration version.
- AppID (Application ID) —the identifier of the GOOSE message.

Within the GSE Control element, there may also be IED Name elements, which specify the devices that will act as subscribers to the GOOSE message. The Name attribute corresponds to GoCBName in the abstract model; Datset is the reference to the data set; and AppID identifies the GOOSE message and corresponds to the GOOSE Control Block in the abstract model.

In Fig. 8.16, the description of the GSE element inside the Communication block in the SCL file is explained. The GSE element is found within ConnectedAP, which describes the device port's connection to the communication network (represented by the Subnetwork element). Each Subnetwork element corresponds to a logically or physically isolated network segment within the overall communication system of the power facility. For example, if there are several separate local networks based on voltage classes, each would be represented by a separate Subnetwork element in the SCL file, describing the network. The connection of devices to this network would be modelled through the appearance of the ConnectedAP element within the relevant Subnetwork element.

```
<Communication>
  ...
  <SubNetwork name="SubNetwork3" type="8-MMS">
    <ConnectedAP iedName="ALL_S001_A3_BC_" apName="AP3">
      ...
      <GSE ldInst="LD1" cbName="gcbA">
        <Address>
          <P type="APPID">0001</P>
          <P type="MAC-Address">01-0C-CD-01-01-01</P>
          <P type="VLAN-ID">06E</P>
          <P type="VLAN-PRIORITY">4</P>
        </Address>
        <MinTime unit="s" multiplier="m">100</MinTime>
        <MaxTime unit="s" multiplier="m">10000</MaxTime>
      </GSE>
      ...
    </ConnectedAP>
      ...
  </SubNetwork>
```

Fig. 8.16 Description of GSE element within communication

The ConnectedAP element has attributes such as iedName (the name of the device) and apName (Access Point Name, representing the communication port connecting the device to the network).

Within ConnectedAP, the GSE element defines that the GOOSE message transmitted by the configured control block will be transmitted through this communication port to the respective subnet.

The GSE element contains attributes like Logical Device Instance (ldInst), which corresponds to the logical device instance that holds the GOOSE Control Block, and cbName, which refers to the name of the control block (as in GoCBName).

Inside the GSE element is the Address element, where communication parameters for the GOOSE message are set, such as VLAN ID, priority, destination MAC address, and Application ID (AppID). The VLAN ID organizes message filtering in the switch, the priority determines the queue for processing packets during network congestion, the MAC address specifies the group (multicast) address, and the AppID is used by manufacturers for identifying and filtering messages at the network interface. The standard does not define strict rules for these parameters, so their assignment must follow local profiles. The Address element also contains MinTime and MaxTime, parameters discussed earlier.

Lastly, it can be considered how GOOSE message subscription is described in another device. As GOOSE messages use multicast transmission, the receiving device must subscribe to these messages, as shown in Fig. 8.17.

Subscription is done using ExtRef elements, located inside Inputs, which are within the logical nodes used for the subscription (either LN or LN0). The ExtRef element is a virtual input for the device, with attributes indicating which data it receives. The identifier for this input is intAddr, and the complete path for the data reference must be specified, including the device name (iedName), logical device ID

```
<LN lnClass="XCBR" lnType="XCBR1" inst="1" prefix="">
  <Inputs>
    <ExtRef intAddr="Tr" desc="Trip" iedName="IED2_PTOC" ldInst="GenericIO" lnClass="PTRC" lnInst="1"
    doName="Tr" daName="general" serviceType="GOOSE" srcLDInst="GenericIO" srcLNClass="LLN0" srcCBName="GoCB" />
    <ExtRef intAddr="Op" desc="Operate" iedName="IED1_XCBR" ldInst="GenericIO" lnClass="CSWI" lnInst="1"
    doName="Pos" daName="opOk" serviceType="Poll" />
  </Inputs>
</LN>
```

Fig. 8.17 Description of GOOSE subscription

(ldInst), logical node path (prefix, lnClass, lnInst), data object (doName), and data attribute (daName), if relevant. Additionally, to ensure clarity about how data enters the virtual input (whether through GOOSE or Report), the sourceCbName attribute, indicating the control block responsible for the signal, is also required.

Thus, a full cycle is completed: firstly, GOOSE message transmission is configured followed by the setup of receiving devices to accept signals from the GOOSE message. The SCL configuration format is standard, universal, and formalized for both GOOSE message transmission and subscriptions. This means that the configuration, once written in SCL, can be interpreted by other users and, more importantly, by software from different manufacturers. This configuration can then be visually represented, tabulated, or otherwise used to generate project documentation.

8.3.3 Mapping the GOOSE Abstract Model to a Specific Communication Protocol

The abstract service SSendGooseMessage, described in IEC 61850 Sect. 7.2, is mapped to the 802.3 (Ethernet) protocol, as outlined in Sect. 8.1 of the standard.

It is important to highlight a specific feature: generally, Sect. 8.1 describes the mapping of most abstract services, and nearly all of them are mapped to the MMS protocol. However, the only service that is mapped to the data link layer of the OSI seven-layer model is the SendGooseMessage service, which translates to "GOOSE message transmission", This approach is taken because transitioning to a higher layer (from physical to data link, from data link to network, transport, session, etc.) results in an increase in the time required to process the information that is encapsulated at lower data levels. Therefore, mapping the SendGooseMessage service directly to the data link layer (Ethernet) maximizes the speed of the protocol and minimizes potential delays. It can be said that this service is assigned to the lowest possible layer, just above the physical level, which directly handles the bits transmitted over various media, whether fibre optics or twisted pair cables.

Mapping GOOSE to the data link layer essentially creates a new type of Ethernet protocol. This is the GOOSE protocol, for which the EtherType identifier 88B8 is reserved. This protocol specifies that it is transmitted in multicast mode, and the destination MAC address follows the format 01-0C-CD-01-XX-YY, where "XX" and "YY" are two configurable parameters. The following details should be noted:

firstly, the first octet "01" indicates that this is a multicast address. A switch receiving an Ethernet frame with a destination MAC address of "01" already understands that if this address is not contained in any filtering tables, the frame should be forwarded to all ports. Consequently, communication occurs in a multicast manner, and the frame is transmitted to all ports, not just to the port connected to a specific device. The next octets 0C-CD are reserved by Technical Committee 57 of the International Electrotechnical Commission (IEC) for its protocols. Therefore, for all protocols defined by IEC TC 57, multicast addresses in the range 0C-CD will be used. The following octet "01" is reserved for GOOSE messages. It will later be shown that for Sampled Values messages, the frame has the same address format, but instead of "01", it will use "04" (this range is reserved for SV). The next two octets can be used to identify messages. Strictly speaking, "XX" in this identifier format is not entirely free for configuration. According to the provisions of the IEC 61850 standard, it must take the value either "00" or "01". However, in practice, this restriction is often relaxed, and both XX and YY are typically configured with free hexadecimal values, making configuration more convenient and flexible from a practical perspective.

Further, the structure of the data block in transmitted GOOSE messages should be considered, as shown in Fig. 8.18. The first element visible in a GOOSE message (i.e., within the Ethernet frame containing the message) is a reference to the GOOSE Control Block in the device's information model. This is the GoCBRef attribute.

The next attribute is Time Allowed to Live (TAL), a crucial attribute related to the message retransmission timing. The standard recommends that the TAL value should be twice the retransmission interval. For example, if a GOOSE message is retransmitted every second in normal mode, then the TAL will be 2 s. If the minimum retransmission time after a data change is set to 3 ms, then the TAL will be 6 ms. This TAL parameter allows the receiving device to know how long it should expect a retransmission of this message, and if the message is not received within the TAL, the device can signal a communication failure.

№	Parameter	Description
1	gocbRef	Reference to the GOOSE control block associated with the publication
2	timeAllowedtoLive	The maximum wait time before the next retransmission
3	datSet	Reference to the data set being published in the GOOSE message
4	goID	GOOSE message identification
5	t	Time at which the last state change was detected
6	stNum	State Number; GOOSE message transmission variation counter
7	sqNum	Sequence Number; GOOSE message counter
8	simulation	Attribute indicating that the message is generated by a testing unit. In the first edition of the standard the attribute was named Test
9	confRev	Counter; increases with each change in the data set configuration
10	ndsCom	Needs Commissioning; indicates whether the GOOSE control block configuration is incomplete or incorrect
11	AllData	Contains all data information within the data set

Fig. 8.18 Structure of the GOOSE message data block

The next attribute, DatSet, is again a reference to the DATASET in the device's information model, whose data is transmitted in this GOOSE message. The following attribute is GoID, the user-defined identifier of the GOOSE message. The fifth attribute is t (timestamp), which corresponds to a change in the StNum counter. StNum is a counter that indicates the number of state changes of the signals transmitted in the message. This means that every time one of the signals changes (e.g., a protection trip occurs and a signal changes), the StNum counter is incremented by one. At this precise moment, the t attribute (timestamp) is also updated. Even if subsequent frames are retransmitted with a different interval, the t and StNum attributes will not change until the next signal state change.

The next attribute, SqNum, represents the sequence number of the packet since the last change in state. Each time a retransmission occurs without any state change, SqNum is incremented by one. However, when the state changes, SqNum is reset, takes on the value of 1, and starts counting again.

The next attribute, Simulation, is also highly important. It is intended to identify messages sent by a test set. It is crucial to be careful here because in the first edition of the IEC 61850 standard, this attribute was called Test, and its interpretation was often incorrect. It was assumed that this attribute was set when a device was switched to test mode. In fact, it should be understood that the device itself should never generate data with the Simulation attribute. This attribute is meant specifically for messages generated by test sets during testing phases, such as when we are testing a protection scheme and simulating signals from a test set bypassing the actual devices.

The next attribute, configuration revision (ConfRev), corresponds to the configuration version and is monitored by the receiving device. It indicates whether the receiving device is receiving data with the same configuration version it is set to process.

The final attribute, NdsCom (needs commissioning), signals the need for reconfiguration. It may be set to true when an empty dataset is transmitted. If a GOOSE message has been configured, but the configuration of the data set was overlooked, the NdsCom attribute should be set, and responsibility for setting this attribute lies with the device manufacturer. Finally, at the end of the message, the actual data defined by the data set is transmitted.

It is worth noting the specific features of mapping the GOOSE model to the Ethernet data link layer. Firstly, at the data link layer, data is addressed using physical addresses, i.e., MAC addresses, rather than IP addresses. It is important to understand that GOOSE, as a subset of Ethernet protocol, operates below the TCP/IP stack. There are no IP addresses in GOOSE messages. As a result, the routing mechanisms typically used by TCP/IP protocols do not work with GOOSE frames. In other words, GOOSE messages cannot be transmitted between different networks using standard router capabilities without additional proprietary configurations; they can only be transmitted within a single local network. Some router manufacturers offer solutions that allow data link layer protocols to be routed by encapsulating these frames into IP packets generated by the router. These packets are then transmitted and unpacked on the receiving end. However, these are proprietary solutions from specific manufacturers and are not standardized.

Packet switching at the data link layer is handled by switches, often referred to as "Layer 2" or Layer 2 switches. It is important to distinguish between switches and routers. Routers are generally considered Layer 3 switches or Layer 3 devices, which handle Layer 3 packets (TCP/IP packets) and provide IP routing. Since IP routing is not possible with GOOSE, packet switching is managed by Layer 2 switches.

Another crucial point to highlight is that Layer 2 switches generally operate at a low hardware level. This means that packet switching and determining the route within the switch (i.e., which port the frame should go to) is performed directly on the chip, i.e., the switching circuitry. As a result, maximum data transmission speed is achieved with minimal delay. These delays are on the order of microseconds, and for short Ethernet frames, they may not exceed 5 μs. Performance is determined more by the physical limitations of the transmission medium than by the capacity of the switching matrix itself. Thus, the delays in GOOSE message transmission are, firstly, close to deterministic, and secondly, they are significantly lower than those experienced when transmitting signals using relay-contact circuits. This is because relay-contact circuits typically include special filtering for input signals or mechanisms to filter "contact bounce". Since there is no "contact bounce" in packet switching, GOOSE message transmission can be several milliseconds faster—up to 10 ms faster—than signal transmission using relays. This has been proven through extensive experimental testing.

When discussing the speed requirements for GOOSE messages, several key points should be noted. Firstly, speed requirements are not applied directly to the GOOSE protocol itself but to the messages that transmit information between specific functions.

These requirements are defined in Chap. 5 of the standard. For signals transmitted between protection and control functions (PAC), there are two types of messages: Type 1A, which transmits breaker trip signals from protection relays, and Type 1B, which transmits other types of signals. The standard defines the concept of Transfer Time, which can be compared to the transmission of signals in classical relay-contact circuits. For each message type, a performance class is assigned, along with the maximum allowable transfer time. In accordance with Sect. 8.1, Type 1A and Type 1B messages are transmitted using the GOOSE protocol, meaning that the speed requirements for these messages are applicable to GOOSE messages.

It is important to understand that the Transfer Time encompasses not only the time it takes for a message to traverse the communication network but also the processing time within devices. Message processing time before sending and after receiving is specific to each device and can be verified during compliance testing. For this, the GOOSE Performance Testing section is included in the test program.

Regarding message transfer time, contrary to the common belief that network switches introduce significant delays, the actual delay caused by the network is one to two orders of magnitude smaller than the processing time within the devices. For example, the maximum delay introduced by a switch when transmitting a message under normal conditions (without queuing) can be estimated to be around 10 μs, and in the presence of queues, it may reach up to 100 μs. Experiments have shown that delays in a normally functioning network (not overloaded with messages) do not exceed a few hundred microseconds. Thus, the network transmission time can be considered negligible compared to the processing time of messages before sending and after receiving, both of which must meet performance requirements during equipment testing. However, it is essential to note that this assumption holds true only if the network is correctly designed and the network equipment is properly configured.

The requirements for transmission time are standardized for each individual class. For message type 1A, two speed classes are defined—P1, P2–, and their corresponding transmission time classes are TT6 and TT5, which should not exceed 3 ms and 10 ms, respectively.

For messages of type 1B, there is one speed class, P3, with transmission time class TT4, which should not exceed one period of industrial frequency—20 ms for 50 Hz networks or 16.67 ms for 60 Hz networks.

As a result, with a correctly designed and configured LAN, and with the use of equipment that meets the performance requirements for the corresponding class, the data transmission time between functions will fall within the specified limits and may not exceed 3 ms. However, IEC 61850 does not explicitly define which messages (signals) should use which classes; these definitions must be established in local normative documents or standard profiles.

The application scope of GOOSE messages is quite broad and is not limited to protection and transmission functions, such as sending breaker trip commands from a protection system. Nevertheless, the primary application is in tasks where fast data transmission from one device to another (or a group of devices) is critical. The most typical use cases are for distributed protection and automation functions, such as busbar protection logic, signal transmission for load shedding, implementation of automatic reclosing functions, as well as breaker tripping from protection systems. This is especially important in substations where a signal can be transmitted from a protection relay terminal, located in the control room, to a breaker in digital form via a discrete signal conversion device (DSCD) installed near the breaker in the switchyard. The protection terminal sends the signal to the DSCD in the form of a GOOSE message, which can be more reliable and faster than using copper wires.

Additionally, GOOSE messages are used to implement operational interlocking of switching devices via connection controllers, which can be located near these devices. However, the use of GOOSE messages is not limited to this list; various other applications are increasingly being found in protection and automation systems.

Finally, the basic structure of a merging unit with GOOSE support must be considered separately. This device is specifically designed for converting discrete signals into GOOSE messages and vice versa (Fig. 8.19). The left side of the device shows the discrete interfaces of the merging unit—discrete inputs and outputs. Through these interfaces, the merging unit connects to primary equipment. For example, the merging unit can be connected to a power breaker, in which case it can receive breaker position and diagnostic circuits. Correspondingly, commands for breaker closing and opening are sent through discrete signals to the breaker's actuator. On the right side of the merging unit, the "IEC 61850 server" is shown schematically. The term "server" here should not be understood as specific server hardware or software, but as a functional block supporting IEC 61850. This block illustrates support for

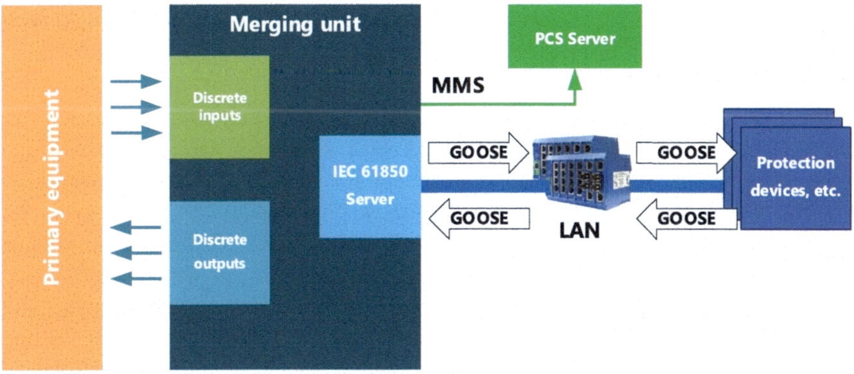

Fig. 8.19 Principal structure of the merging unit

IEC 61850 in terms of GOOSE message reception and transmission. Additionally, the IEC 61850 server may also implement MMS protocol support for transmitting diagnostic information (and possibly discrete input values) directly to the SCADA system, although these are considered additional or optional functions.

It can be assumed that the values of discrete signals arriving at the merging unit input are transmitted as data in a GOOSE message, and conversely, changes in data within incoming GOOSE messages result in changes in the corresponding discrete outputs. The merging unit may also implement logic for processing incoming signals and GOOSE messages, for example, to ensure compliance with the correct semantics of the IEC 61850 standard. This processing is particularly necessary when a single "switchgear position" signal, conforming to a two-point format as per IEC 61850, is formed based on two discrete signals, "Closed Position" and "Open Position."

8.4 Sampled Measured Values (SV) Protocol

8.4.1 Abstract Model, Sampled Values Control Block, and Communication Services

The abstract model of Sampled Values describes the mechanism for transmitting instantaneous sample values from measuring transformers, analogue signal converters, and other devices, without being tied to specific data transmission technologies or communication protocols. The description of this abstract model is provided in Sect. 7.2 of the IEC 61850 standard. The abstract model of Sampled Values outlines the principles for transmitting instantaneous sample values, the required attributes of control blocks, and their names, but does not specify the details of the data transmission technology. The specific protocol for Sampled Values is detailed in Sect. 9.2 of the IEC 61850 standard [9].

As shown in Fig. 8.20, the abstract model of Sampled Values is presented with a connection involving three current measuring transformers. Each of these transformers is connected to analogue signal converters (ASC). In this case, it is not essential whether classical electromagnetic transformers with connected ASCs are used or whether digital current transformers with a digital interface are used. This is not the core element of the abstract model for Sampled Values transmission. However, the following is important: firstly, these physical devices, in this example, act as publishers, and the transmission of Sampled Values, like the transmission of GOOSE messages, is carried out in multicast mode. In this context, there are two roles in which devices communicate with each other: publishers, who provide the information, and subscribers, who receive it. Importantly, in this relationship, the subscribers are aware of the publishers, but the reverse is not true. A publisher does not know who might be subscribed to it [10].

The next point to note about the publishers, or analogue signal converters (ASCs), is that they take data from the analogue circuit and convert it into digital samples with a certain sampling rate. These samples are then formatted, sent to the communication network, transmitted, and, upon receiving these samples, the subscriber processes them. Publishers must be precisely time-synchronized. Indeed, if it is assumed that they are not synchronized accurately, with some error, a situation may arise where the subscriber receives samples published by the publishers at different moments in time. If the subscriber processes them, it could result in incorrect data. For example, a phase mismatch of the samples could cause a differential current, potentially triggering a protection mechanism. To prevent this from happening, publishers must be precisely

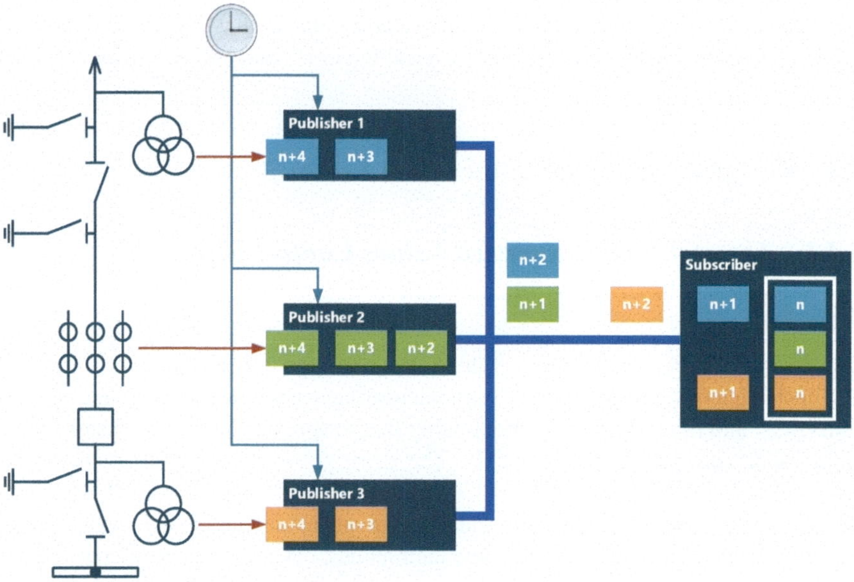

Fig. 8.20 Abstract model of SV transmission

time-synchronized. When processing a sample, each publisher assigns a timestamp that corresponds to the exact moment the sample was taken.

Next, an important requirement for the subscriber is that it must process only those samples taken at the same moment in time. This can be inferred indirectly by looking at the counter number, which must always be transmitted with the samples. The counter number is assigned by the publisher according to the exact moment the sample was taken.

The following are the three basic points for transmission via the SV protocol:

- The publisher transmits Sampled Values in multicast mode.
- All publishers must be time-synchronized.
- The receiving device must take sample numbers into account when processing.

As shown in Fig. 8.20, the samples may not arrive at the subscriber in order. At the moment depicted, samples with number n are being processed, and in the buffer, there are samples with number n + 1 from two devices, but the sample n + 2 is incoming. Therefore, an important requirement for the subscriber is the presence of an input buffer capable of accumulating samples with different numbers and considering these numbers during processing. Otherwise, the sample with number n + 2 could either be lost due to a counter failure or be placed in the wrong order, leading to incorrect data processing. It is also important to note that this out-of-order sample transmission may be caused by the fact that different publishers may generate samples with varying delays. It should be understood that this delay does not occur in the data transmission network (which, in this case, is assumed to contribute no error) but at the point where the samples are generated. For example, certain types of current transformers may introduce delays of up to several hundred microseconds in the sample generation process, while the analogue signal converter might do this much faster, with a shorter delay. A delay of several hundred microseconds, for instance, 500 μs, in the context of 80 samples per period, or with a sampling rate of 4 kHz, accounts for 2 samples. This means that the sample numbers from different devices may differ by 2. Consequently, a requirement arises for the subscriber to have a buffer that can store at least two samples deep in order to synchronize all the samples from the devices it is subscribed to.

When the subscriber receives and processes data from multiple publishers, time synchronization becomes critically important. If the subscriber processes measurements from only one publisher, time synchronization is not as crucial. Consequently, the system can operate without time synchronization. For example, if we consider the connection between a single current transformer and a protection relay device, and focus on the overcurrent protection (OC) function within that device, it becomes clear that this function uses measurements from only one current transformer [11]. Therefore, even if this current transformer is not synchronized with other neighbouring transformers, the overcurrent protection function will still operate correctly. This is an important consideration to avoid placing excessive demands on the synchronization system and its reliability. In other words, we can conclude that even if time synchronization fails in the system due to some factors, it will not affect the operation of backup protection (such as overcurrent protection).

№	Parameter	Description
1	MsvCBName	Name of the control block in the device model
2	MsvCBRef	Multicast Sampled Value Control Block
3	svEna	The attribute of enabling/disabling the control unit
4	MsvID	System wide unique ID of the sending application
5	DatSet	The name of the data set being sent
6	ConfRev	Configuration Revision of the MSV Control Block
7	SmpMod	Sample Mode: samples/period, samples/second, seconds/sample
8	SmpRate	Sample Rate
9	OptFlds	Optional fields in the packet
10	DstAddress	Адрес назначения

Fig. 8.21 Sampled values control block

Next, we turn to the Sampled Values control block (Fig. 8.21)—the Multicast Sampled Values Control Block (MSVCB). The MSVCB itself is completely analogous to the GOOSE message control block. That is, we have a transmitting device, where a dataset is created, then the Sampled Values control block is configured. Among its parameters, a reference to the created dataset is included, and this control block handles the transmission of sampled values to the network [12]. The main parameters of the MSVCB control block are as follows.

Firstly, the block has a name. Similar to the GOOSE Control Block, the name is a fixed parameter set by the manufacturer, but it may be configurable for certain devices. Next, there is a reference to the control block within the device's information model. This parameter is also non-configurable but is important for identifying the block. The next parameter, SvEna, is an attribute that enables or disables the control block. When disabled, the control block does not generate Sampled Values. SvEna is typically available for management via MMS. Alternatively, this parameter may be disabled, in which case SvEna becomes unavailable for management. MsvID is the Sampled Values identifier. It is the main identification parameter that must be assigned in the project. It is recommended to use a specific naming system (SVID) to uniquely identify the corresponding SV stream. The fifth attribute, DatSet, is a reference to the data set in the device model. The sixth attribute, ConfRev, similarly to GOOSE messages, is used to inform receivers about the configuration version currently loaded into the sender of the corresponding Sampled Values packet. If the version does not match the configuration version initially loaded into the receiver, the receiver will issue a corresponding alarm. SmpMod is a user-defined attribute that sets the sampling frequency mode, i.e., it determines the following parameter, SmpRate. There are three options: samples per industrial frequency period (this is the default parameter), samples per second, or intervals in seconds per sample.

The next parameter, SmpRate, which is directly related to SmpMod, defines the sampling rate. For example, in the first edition of the standard, SmpMod defined samples per industrial frequency period, which is why, in the second edition, this parameter became the default. As a result, SmpRate is set as 80 or 256 samples. To calculate the number of samples per second, the industrial frequency must be multiplied by the number of samples. For instance, multiplying 80 samples by a

50 Hz industrial frequency results in 4,000 samples per second (or a 4 kHz sampling rate). This method of setting the sampling rate is not very convenient as it requires a second parameter to uniquely identify the sampling frequency. The second parameter is the industrial frequency. As practice has shown, issues with subscribing to Sampled Values often arise due to discrepancies in default frequencies (50 or 60 Hz), and it is not always possible to clearly identify the frequency at which the samples are transmitted. Therefore, the SmpMod parameter was introduced with the options mentioned above. Now, the sampling mode can be set in samples per second, and in this case, the sampling rate is specified directly (e.g., 4 kHz), without relying on the industrial frequency. The next attribute, Optional Fields, refers to optional fields that may be included in the frame transmitted over the network. The tenth attribute, Destination Address, is the destination address of the packet, i.e., the group MAC address.

The standard defines the following services for the Multicast Sampled Values Control Block:

(1) Send Multicast Sampled Values message (SendMSVmessage)—transmission of sampled instantaneous values;
(2) GetMSVCBValues—reading control block parameters;
(3) SetMSVLCBValues—setting or writing control block parameters.

The SendMSVmessage service is used to transmit the data (samples) generated by an analog signal converter or electronic current or voltage transformer with a digital interface. The other two services are used for reading parameters. For example, a client system or SCADA system can read control block parameters or set them remotely. For instance, the client system can disable the transmission of Sampled Values if necessary. Alternatively, certain parameters can be changed and the transmission can be resumed. This is managed by the GetMSVCBValues and SetMSVCB-Values services. It is important to note that these services may not be implemented in all devices. For example, in the case of an analog signal converter, it may not support the communication services GetMSVCBValues and SetMSVCBValues. In such cases, these parameters can only be changed via device configuration, not through the client system. The converter only outputs sampled instantaneous values, and to modify its configuration, corresponding software for the device must be used.

8.4.2 Mapping the Sampled Values Abstract Model to a Specific Communication Protocol

Next, we can examine the mapping to a specific communication protocol, as described in Sect. 9.2 of the IEC 61850 standard. There are three communication services, of which only one—SendMSVmessage—is mapped to the Ethernet protocol in accordance with the provisions of Sect. 9.2 [13]. The remaining services are mapped to the MMS protocol. That is, reading and controlling the control block is done via the MMS protocol, which has no relation to Sect. 9.2. Similarly to the GOOSE protocol, a separate EtherType is reserved for the Sampled Values protocol, which is very similar to GOOSE (GOOSE's EtherType is 88B8, and for Sampled Values it is 88BA). It is executed in multicast mode (Multicast broadcasting), where the destination MAC address is specified in the following format. As with GOOSE messages, the first three octets are the same—01, 0C, CD. This is the multicast range of MAC addresses reserved by IEC Technical Committee 57. The next octet, "04", is the range reserved for SV messages (the corresponding octet in GOOSE was "01"). Then "xx-yy" are message identifiers that can be used within a project. These are hexadecimal values that the user can define according to their project. Similar to GOOSE messages, in SV streams, the values of "xx" according to IEC 61850 must be either 00 or 01. However, in practice, there are projects that require more address values. In Sampled Values, this is not yet as critical, but there are projects where more than 500 GOOSE messages are encountered. In such cases, the range of 00 or 01 for the corresponding octet may be insufficient, so this value can be varied.

The Sampled Values protocol, mapped to the Ethernet communication protocol, i.e., to the data link layer, is essentially assigned to the lowest level of the OSI seven-layer communication stack. This minimizes delays in data transmission. Therefore, it has an extremely high speed, much higher than all TCP/IP protocols, which are based on transport and higher layers [14].

For a better understanding of what the data transmitted via the SV protocol consists of, it is necessary to examine in more detail the structure of the Ethernet frame for this protocol (Fig. 8.22).

Any Ethernet frame, whether it is SV, GOOSE, or any other Ethernet-based protocol (including IP protocols), has the same structure, which includes the destination MAC address, the sender's MAC address, followed by a protocol type field indicating the protocol contained within the frame, and the total frame length. After these fields, the data carried by the protocol in the frame, in this case, the SV protocol, is included, and this field is called the savPdu.

The protocol data block (savPdu) has a rather simple structure. At the beginning of the data block is the noASDU field, which indicates the number of ASDU blocks

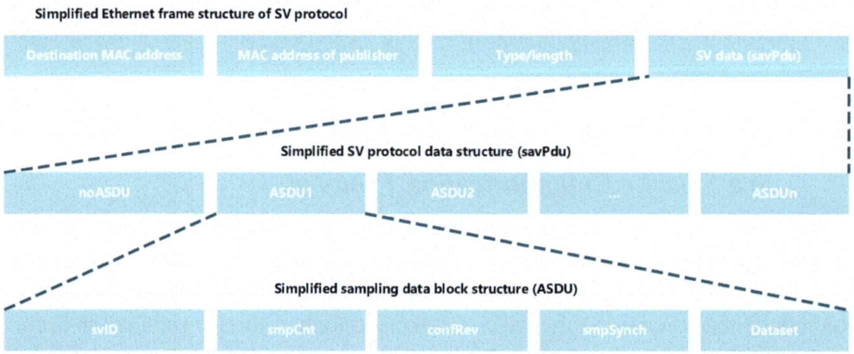

Fig. 8.22 Ethernet frame structure for SV

transmitted in the frame. ASDU is essentially a sample, meaning that a frame can contain one or more samples. The number of samples is defined by the noASDU parameter. Next, it is necessary to consider the structure of one ASDU to understand the data being transmitted in a sample. The first field is the SV stream identifier—svID. The svID parameter is a configurable parameter that is set during device configuration. The svID must uniquely identify each SV stream. The next parameter, smpCnt, is a cyclic sample counter that resets every second, and accordingly, it counts from 0 to (f-1), where f is the sampling frequency in hertz. The next field, confRev, indicates the version number of the configuration for the respective SV transmission block. The smpSynch flag indicates the synchronization state of the sending device at the time of forming the respective sample. At the end of each sample, i.e., each ASDU, there is a data set transmitted as part of the sample. The composition of the data set is a configurable parameter. However, according to the IEC 61850-9-2LE specification, the data set includes 8 analogue values (4 currents and 4 voltages) and their corresponding quality flags, meaning a total of 16 attributes.

When considering the SV protocol, it should be noted that there is an international standard, IEC 61850-9-2, which describes the mapping of communication services to a specific Ethernet communication protocol in accordance with Sect. 7.2. In particular, the mapping of the SendMSVmessage communication service, as mentioned earlier. At the same time, Sect. 9.2 provides greater freedom for the practical implementation of the communication protocol. For example, regarding the sampling frequency, Sect. 9.2 does not impose any restrictions on the sampling frequencies used. Additionally, the standard allows for the packaging of multiple samples into a single frame. For instance, if a device samples at a frequency of 4 kHz, it could generate 4,000 frames per second. However, instead of sending a frame with each sample, it packs, for example, 4 samples into each frame. This way, samples taken at 4 kHz are transmitted into the communication channel at only 1 kHz. As a result, each frame carries 4 samples, which the receiver processes sequentially.

8.4.3 Configuration of Sampled Values Transmission in SCL Language

The parameterization of data transmission according to IEC 61850 is implemented in the same ideology as the parameterization of data transmission using the GOOSE message transfer service. Firstly, a data set is formed, which will be transmitted, then the control block for transmitting Sampled Values is parameterized. In the data set shown in Fig. 8.23, the format of the data set according to IEC 61850-9.2LE can be seen. It consists of 8 measurements (4 currents and 4 voltages) and 8 quality flags, each corresponding to its respective measurement. Each current and voltage channel corresponds to a separate logical node, and in accordance with IEC 61850-9.2LE, they have a specific naming structure. The prefix indicates the winding number and phase (e.g., I01A, I01B, I01C) and the instance number, i.e., "lnInst", which

```
<DataSet name="PhsMeas1" desc="IEC 61869-9 9-2LE dataset for SV streams">
    <FCDA ldInst="MU01" prefix="I01A" lnClass="TCTR" lnInst="1" doName="AmpSv" daName="instMag.i" fc="MX"/>
    <FCDA ldInst="MU01" prefix="I01A" lnClass="TCTR" lnInst="1" doName="AmpSv" daName="q" fc="MX"/>
    <FCDA ldInst="MU01" prefix="I01B" lnClass="TCTR" lnInst="2" doName="AmpSv" daName="instMag.i" fc="MX"/>
    <FCDA ldInst="MU01" prefix="I01B" lnClass="TCTR" lnInst="2" doName="AmpSv" daName="q" fc="MX"/>
    <FCDA ldInst="MU01" prefix="I01C" lnClass="TCTR" lnInst="3" doName="AmpSv" daName="instMag.i" fc="MX"/>
    <FCDA ldInst="MU01" prefix="I01C" lnClass="TCTR" lnInst="3" doName="AmpSv" daName="q" fc="MX"/>
    <FCDA ldInst="MU01" prefix="I01N" lnClass="TCTR" lnInst="4" doName="AmpSv" daName="instMag.i" fc="MX"/>
    <FCDA ldInst="MU01" prefix="I01N" lnClass="TCTR" lnInst="4" doName="AmpSv" daName="q" fc="MX"/>
    <FCDA ldInst="MU01" prefix="U01A" lnClass="TVTR" lnInst="1" doName="VolSv" daName="instMag.i" fc="MX"/>
    <FCDA ldInst="MU01" prefix="U01A" lnClass="TVTR" lnInst="1" doName="VolSv" daName="q" fc="MX"/>
    <FCDA ldInst="MU01" prefix="U01B" lnClass="TVTR" lnInst="2" doName="VolSv" daName="instMag.i" fc="MX"/>
    <FCDA ldInst="MU01" prefix="U01B" lnClass="TVTR" lnInst="2" doName="VolSv" daName="q" fc="MX"/>
    <FCDA ldInst="MU01" prefix="U01C" lnClass="TVTR" lnInst="3" doName="VolSv" daName="instMag.i" fc="MX"/>
    <FCDA ldInst="MU01" prefix="U01C" lnClass="TVTR" lnInst="3" doName="VolSv" daName="q" fc="MX"/>
    <FCDA ldInst="MU01" prefix="U01N" lnClass="TVTR" lnInst="4" doName="VolSv" daName="instMag.i" fc="MX"/>
    <FCDA ldInst="MU01" prefix="U01N" lnClass="TVTR" lnInst="4" doName="VolSv" daName="q" fc="MX"/>
</DataSet>
```

Fig. 8.23 SV DataSet

corresponds to the phase number in sequence (A, B, C—1, 2, 3, and 4—for neutral). The same applies to voltages. According to IEC 61850-9.2LE, the logical device that outputs instantaneous values must have a name in the MU format, followed by the measurement point. In this case, the point is "01". Thus, MU01 I01A TCTR1, followed by the data object, either Amp for TCTR or Vol for TVTR, and then the data attribute InstMag.i.

The description of the Sampled Values (SV) control block is performed similarly to the control block for GOOSE message transmission. As shown in Fig. 8.24, the block name is indicated, in this case, "MSVCB01," referencing the "DATA-SET" (the name of the DATA-SET, according to IEC 61850-9.2LE, is fixed and must be named "PhsMeas1"). The next attribute is ConfRev (fixed according to IEC 61850-9.2LE, with a value of one). In the case of applying a different profile, the configuration may be modified. The next attribute is SMVID, the application identifier for the Sampled Values stream used in the project. The next parameters are Sample Rate and the number of ASDU in the packet. The Sample Rate is specified as 80 samples per period, and the number of ASDU is one.

The next section, where the transmission of Sampled Values is configured, is Communications. Inside the corresponding subnet, within the Subnetwork element, there is the Connected AP element, which corresponds to the access point of our device. That is, the IED Name attribute in the Connected AP must match the Name

```
<ConnectedAP iedName="RTDS_GTNETx2_SV" apName="P1">
        <Address>
            ...
            <P type="IP">172.24.9.183</P>
            <P type="IP-TCP-PORT">102</P>
        </Address>
    <SMV ldInst="MU01" cbName="MSVCB01" desc="9-2LE SV stream (F4000S1I4U4 and F4800S1I4U4)">
        <Address>
            <P type="VLAN-ID">075</P>
            <P type="VLAN-PRIORITY">4</P>
            <P type="MAC-Address">01-0C-CD-04-00-22</P>
            <P type="APPID">4011</P>
        </Address>
    </SMV>
</ConnectedAP>

<IED name="RTDS_GTNETx2_SV" manufacturer="RTDS TECHNOLOGIES INC." >
    ...
                <LN0 lnType="LN0" lnClass="LLN0" inst="">
                    ...
                    <SampledValueControl name="MSVCB01" datSet="PhsMeas1" smvID="RTDSMU0101" smpRate="80" smpMod="SmpPerPeriod" nofASDU="1" multicast="true" confRev="1">
                        <SmvOpts refreshTime="false" sampleSynchronized="true" sampleRate="false" dataSet="false" security="false" />
                    </SampledValueControl>
                </LN0>
```

Fig. 8.24 Description of the sampled values control block

```
<Inputs>
  <ExtRef srcCBName="MSVCB01" iedName="QT2G_ORU2_02A1_A" ldInst="MU01" prefix="I01A" lnClass="TCTR"
  lnInst="1" doName="Amp" daName="instMag.i" intAddr="W2G_2P_A1_LPLD/MMXU1.A.phsA.cVal.mag.f" />
  <ExtRef srcCBName="MSVCB01" iedName="QT2G_ORU2_02A1_A" ldInst="MU01" prefix="I01B" lnClass="TCTR"
  lnInst="2" doName="Amp" daName="instMag.i" intAddr="W2G_2P_A1_LPLD/MMXU1.A.phsB.cVal.mag.f" />
  <ExtRef srcCBName="MSVCB01" iedName="QT2G_ORU2_02A1_A" ldInst="MU01" prefix="I01C" lnClass="TCTR"
  lnInst="3" doName="Amp" daName="instMag.i" intAddr="W2G_2P_A1_LPLD/MMXU1.A.phsC.cVal.mag.f" />
  <ExtRef srcCBName="MSVCB01" iedName="QT2G_ORU2_02A1_A" ldInst="MU01" prefix="I01N" lnClass="TCTR"
  lnInst="4" doName="Amp" daName="instMag.i" intAddr="W2G_2P_A1_LPLD/MMXU1.A.neut.cVal.mag.f" />
  <ExtRef srcCBName="MSVCB01" iedName="QT2G_ORU2_02A1_A" ldInst="MU01" prefix="U01A" lnClass="TVTR"
  lnInst="1" doName="Vol" daName="instMag.i" intAddr="W2G_2P_A1_LPLD/MMXU1.PhV.phsA.cVal.mag.f" />
  <ExtRef srcCBName="MSVCB01" iedName="QT2G_ORU2_02A1_A" ldInst="MU01" prefix="U01B" lnClass="TVTR"
  lnInst="2" doName="Vol" daName="instMag.i" intAddr="W2G_2P_A1_LPLD/MMXU1.PhV.phsB.cVal.mag.f" />
  <ExtRef srcCBName="MSVCB01" iedName="QT2G_ORU2_02A1_A" ldInst="MU01" prefix="U01C" lnClass="TVTR"
  lnInst="3" doName="Vol" daName="instMag.i" intAddr="W2G_2P_A1_LPLD/MMXU1.PhV.phsC.cVal.mag.f" />
  <ExtRef srcCBName="MSVCB01" iedName="QT2G_ORU2_02A1_A" ldInst="MU01" prefix="U01N" lnClass="TVTR"
  lnInst="4" doName="Vol" daName="instMag.i" intAddr="W2G_2P_A1_LPLD/MMXU1.PhV.neut.cVal.mag.f" />
</Inputs>
```

Fig. 8.25 Description of subscription to sampled values

attribute of the IED element, where the Sampled Values control block is configured. Next, inside the Connected AP element, an SMV element is provided, which refers to the control block (in this case, MSVCB01, located in MU01). Then, the communication parameters are defined. Inside the Address section, VLAN ID, VLAN Priority, MAC address, and Application ID are specified.

The description of the subscription to Sampled Values, just like with GOOSE messages, is performed using ExtRef elements, which are located within the Inputs section of any logical node (Fig. 8.25). At the same time, the IEC 61850 standard does not specify in which exact logical node the subscriptions should be made. This decision is left to the manufacturer.

Inside the ExtRef element, a reference to the control block Source CB Name is specified, which corresponds to the name of the device from which the data is received. Next, the full reference to the data to be transmitted is provided, including the Logical Device Instance, prefix, class, etc., as well as the name of the internal address to which these measurements will be assigned. Similarly, all other subscriptions are described in the same format. Thus, the subscribing device receives information about which Sampled Values streams it should subscribe to.

From a practical perspective, it is also useful to have information about the network load caused by the transmission of specific data streams [15]. By calculation, it can be determined that the load from a single SV stream with a sampling frequency of 80 samples per period at a 50 Hz industrial frequency is 5.6 Mbps. Accordingly, the load from a 256-stream is approximately 13.1 Mbps. This is the maximum value considering the maximum length of fields whose length may vary. Experimental results confirm that the maximum load on a 100 Mbps interface is around 20 streams at 80 samples per period with a system frequency of 50 Hz. This number of streams (twenty) can be transmitted without frame loss through the switches on which the tests were conducted (from various manufacturers). However, if more than 20 streams are input into the switch, the bandwidth capacity is exceeded, and some frames begin to be lost.

Control Questions and Tasks

1. What data transmission protocols are used in digital substations according to IEC 61850? For what purposes is each protocol used?
2. Explain what client–server communications are. What procedures are possible in client–server communication?
3. Which protocol is used to integrate devices into the SCADA system of a digital substation? Which standard describes this protocol? Explain what types of reports exist on digital substations within the framework of IEC 61850.
4. What is the GOOSE protocol used for? Briefly explain its general characteristics.
5. At which level of the OSI model is the GOOSE protocol located? What is the reason for this? How is the transmission of GOOSE messages carried out? How is message delivery ensured, and how is the transmission channel status monitored?
6. What is the Sampled Values protocol used for? Briefly explain its general characteristics. Why is time synchronization necessary for the Sampled Values protocol on intelligent electronic devices in digital substations?
7. What was the need for the introduction of IEC 61850-9-2LE? What is the sampling frequency of analogue signals, the structure of data sets, and the regulated method of time synchronization according to IEC 61850-9-2LE?

References

1. Humeres Flores P (2023) Planning and design for IEC 61850 implementation. IEC 61850 principles and applications to electric power systems. Springer, pp 167–188
2. Karantaev V, Latyshov K, Karpenko V, Kiselnikov S (2023) Particular issues of implementation of embedded cryptographic instruments of information protection digital substation devices exchanging messages IEC 61850–8–1 Protocol (MMS). 2023 6th International scientific and technical conference on relay protection and automation (RPA), pp 1–19
3. Aftab M, Hussain S, Ali I, Ustun T (2020) IEC 61850 based substation automation system: a survey. Int J Electr Power Energy Syst 120:106008
4. Ren Y, Cao F, Qing L, Yang Q (2000) Research on MMS and Its application in the communication protocol of electric power system. Autom Electr Power Syst 24(19):66–69
5. Lahza H, Radke K, Foo E (2018) Applying domain-specific knowledge to construct features for detecting distributed denial-of-service attacks on the GOOSE and MMS protocols. Int J Crit Infrastruct Prot 20:48–67
6. Kriger C, Behardien S, Retonda-Modiya J (2013) Detailed analysis of the GOOSE message structure in an IEC 61850 standard-based substation automation system. Int J Comput Commun Control 8(5):708–721
7. Reda H, Ray B, Peidaee P, Anwar A et al (2021) Vulnerability and impact analysis of the IEC 61850 goose protocol in the smart grid. Sensors 21(4):1–20
8. Cokić A, Čeljo A (2019) Implementation of protection and control systems in the transmission SS 110/10(20)/10 kV Using IEC 61850 GOOSE messages. Adv Technol Syst Appl III:243–254
9. Kumar S, Abu-Siada A, Das N, Islam S (2021) Toward a substation automation system based on IEC 61850. Electronics 10(3):310
10. Ingram D, Schaub P, Taylor R, Campbell D (2013) Performance analysis of IEC 61850 sampled value process bus networks. IEEE Trans Industr Inf 9(3):1445–1454

11. Penthong T, Ginocchi M, Ahmadifar A, Ponci F et al (2023) IEC 61850-based protection scheme for multiple feeder faults and hardware-in-the-loop platform for interoperability testing. IEEE Access 11:65181–65196
12. Melo A, Roldan-Fernandez J, Gonzalez-Longatt F, Burgos-Payan M et al (2024) Novel IEC 61850-based off-site engineering and validation methodology for protection, automation, and control systems. Electric Power Syst Res 232:110409
13. Zakonjšek J (2023) CT/VT sampled value acquisition applied to IEC 61850. IEC 61850 principles and applications to electric power systems. Springer, pp 301–319
14. El Hariri M, Harmon E, Youssef T, Saleh M et al (2019) The IEC 61850 sampled measured values protocol: analysis, threat identification, and feasibility of using NN forecasters to detect spoofed packets. Energies 12:3731
15. Qing C, Hong-bin L, Ben-xiong H (2010) An innovative combined electronic instrument transformer applied in high voltage lines. Measurement 43(7):960–965

Chapter 9
Information Security of the Digital Substation

9.1 Key Provisions

Modern power systems (PS), referred to as traditional systems, began developing in the early twentieth century. Initially, power plants were connected solely to local loads, primarily lighting and electric transportation systems [1, 2]. Consumers did not expect highly reliable power supplies, so planned and emergency outages were often accepted as inevitable and not catastrophic. As society's dependence on electricity grew, so did the need to enhance the reliability of its supply. Systems connecting multiple generating stations and transmission lines were created, enabling redundancy, and higher voltage levels were used for long-distance power transmission. The nodes where transmission lines interconnected or voltage levels changed were termed "substations." Substations frequently utilized protective devices to isolate faults within the power system, preventing failures from compromising the entire system. Operating personnel were often stationed at these crucial nodes in the electrical system to monitor and respond swiftly to emerging issues. Communication with central dispatch was maintained through any available means (often by telephone), allowing personnel to report the system's status. Such "manned" substations remained standard throughout the first half of the twentieth century.

Modern PSs are frequently described as "the largest machine created by humanity" due to their vast physical size (spanning hundreds to thousands of kilometres) and synchronized operations. Ensuring the reliability of such a complex, large-scale system, and preventing it from disintegrating, is a challenging task. Substations play a crucial role in the operation of PSs. The high reliability of today's PSs relies, among other factors, on technologies known as Supervisory Control and Data Acquisition (SCADA), which allow for remote monitoring and control of key system parameters.

Following IEC 61850 recommendations and adopting SCADA systems has led to the development of digital substations (DSs), which are becoming increasingly complex and interconnected due to advancements in information and communication technology. This increased complexity and interconnectedness of SCADA systems

V. Y. Ushakov et al., *Digitalization of Electrical Power Engineering*, Power Systems,
https://doi.org/10.1007/978-3-031-95705-5_9

has introduced a wide range of cybersecurity threats. These threats originate not only from external factors, such as terrorists, hackers, or competitors, but also from internal sources, such as former employees, dissatisfied with termination, or technical staff, unhappy with working conditions. In practice, an attacker can gain unauthorized cyber access to SCADA systems, exploit vulnerabilities, and conduct attacks with severe consequences, including data breaches, operational disruptions, or even catastrophic physical damage.

In recent years, malicious cybersecurity incidents have occurred at industrial facilities worldwide. For example, in July 2010, the Stuxnet worm attacked Iranian nuclear facilities, becoming one of the most notable cyberattacks directly impacting industrial infrastructure. In December 2015, a coordinated cyberattack utilizing the BlackEnergy malware caused a power outage affecting at least 80,000 customers in western Ukraine, marking the first known power outage induced by a cyberattack. Since then, additional attacks have occurred, including the TRITON/TRISIS malware, which in 2017 targeted facilities using Schneider Electric systems.

The possibility of catastrophic damage to critical infrastructures from future attacks like Stuxnet or BlackEnergy has raised significant concerns among governments globally.

9.2 Cyberattacks and Cybersecurity Threats at Substations

Historically, substation control and protection schemes relied on single-function devices connected via wired communication [1]. Over recent decades, the deployment of multifunctional IEDs has enabled expanded functionality, reducing the number of required devices. Additionally, Local Area Networks (LANs) have replaced traditional wired connections for substation control. The International Electrotechnical Commission has published communication standards, including IEC 61850, which has become the foundation for protection and automation systems within and between substations, as well as between substations and remote facilities such as control centres. However, the increased use of electronic devices exposes secondary systems at substations to malicious interference in the form of cyberattacks or intentional electromagnetic interference.

9.2.1 Intentional Electromagnetic Interference

One aspect of IEC 61850 is its emphasis on compatibility among IEDs from different manufacturers. Another advantage is the publication of an open data dictionary for IEDs and communication services designed to perform most protection and automation functions. Additionally, IEC 61850 and other communication standards include specifications for peer-to-peer communication over high-speed networks using Internet Protocol (IP), which presents significant cybersecurity challenges.

This has necessitated the establishment of requirements to prevent unauthorized cyber access and misuse of IEDs.

Cyberattacks can be categorized into four types as outlined in CIGRE Technical Brochure 603, "Application and Administration of Cybersecurity Measures for Protection and Control Systems".

(1) ***Information-gathering attacks***, including data viewing or alteration at substations, eavesdropping, recording device communications between IEDs and authorized users, and analysing traffic from repeated communication sessions.
(2) ***Spoofing attacks***, such as substitution, cloning, and replay, to simulate legitimate access to substation IEDs or obtain authorized access.
(3) ***Blocking attacks***, targeting the exhaustion of substation IED and network resources or creating communication interference using tactics like denial of service, jamming, and malware.
(4) ***Privacy attacks***, which aim to disclose confidential information about legitimate users or groups of protection and control system users.

Necessary conditions for successful defence against cyberattacks include:

- Clear and unambiguous security policies and procedures for power supply enterprises.
- Reliable security mechanisms provided by substation devices and services.

Cybersecurity threats to substation systems include vulnerabilities introduced during development. Substation engineers often find it challenging to identify and resolve vulnerabilities embedded in protection and control systems' hardware and software by manufacturers.

The most common cybersecurity threat is a vulnerability that enables malicious code to be embedded in system software. This can allow unauthorized access, circumventing access controls established by developers or engineers, such as remote access to protection relay devices or user privilege escalation to administrator level. Typically, administrator privileges include the ability to modify other users' privileges.

Code injection attacks can be implemented by embedding binary or source code. In the first case, malicious code is inserted into a binary program to alter its behaviour. In the second case, it involves interacting with applications in protection and control systems written in interpreted programming languages such as JavaScript, PHP, and SQL. Consequently, this attack type primarily concerns web applications. Common vulnerabilities in this category include cross-site scripting (XSS) and SQL injection. XSS involves inserting malicious JavaScript code into existing web applications, which can then allow any user (or an attacker-specified user) access to the specific application.

9.2.2 *Deployment and Maintenance Platform Vulnerabilities*

One common type of vulnerability involves enabled software services that are either unused or unknown to engineers responsible for protection and control system security. These services are often vulnerable due to a lack of monitoring. An example of an infrequently used service is file sharing, which is enabled by default in modern Windows operating systems. Another example is the File Transfer Protocol (FTP) server, allowing a user to remotely access system file data without the engineer's consent.

9.2.3 *Firewall Configuration Errors*

If we assume that a substation firewall is in the gateway or LAN routers, configuring it correctly is challenging for substation engineers due to the complexity of firewall rules. However, engineers must ensure correct firewall configuration, as its security is assumed to be trustworthy. Verification of firewall configuration accuracy should be conducted during maintenance to achieve this confidence. Regardless of the firewall's location, configuration errors provide attackers with access to vulnerable control and protection system components and their data.

9.2.4 *Online Password Cracking*

For system software that uses passwords, there should be a function limiting the number of permitted password attempts (typically, a maximum of three attempts). Without this feature, attackers can compromise the authentication mechanism to gain access. If the password is weak, the vulnerability is severe, as an attacker may easily guess it.

Sometimes, an attacker can obtain entire databases of substation system user accounts, such as from Active Directory servers. If the information is poorly encrypted or unencrypted, attackers can extract the data from the database. For this reason, control and protection service managers must strictly adhere to cybersecurity rules and relevant regulatory documents.

9.2.5 *Inadequate Access Control Measures*

Poorly configured access control measures can lead to users of the system being given excessive or insufficient privileges, such as granting administrative access to individuals or groups who should only have read-only access. Overly strict access control can also cause issues, such as improper termination of services or the transmission of sensitive credentials to unauthorized personnel. Control and protection system engineers should regularly review who has access control privileges and ensure they are appropriately aligned with the user's role and responsibilities.

9.2.6 Network Traffic Analysis and Manipulation

An attacker with the ability to intercept and record transmitted data may conduct various types of attacks, such as replaying previously sent messages to deceive system operators regarding the state of the power system. It is not difficult to intercept plain text passwords. Merely adding a randomly selected encryption mechanism is insufficient to prevent attackers from eavesdropping and recording message traffic. This applies to both wired and wireless communications, but special attention should be given to wireless communication. For instance, utilities using Wired Equivalent Privacy (WEP) encryption for IEEE 802.11 wireless networks are vulnerable, as WEP encryption mechanisms can be easily compromised. Control and protection system engineers should review all remote wireless access to the automation system, avoiding wireless access that relies on WEP encryption as it is unreliable for interacting with protection and control systems.

9.3 Ensuring Cybersecurity

9.3.1 Practical Cybersecurity Solutions

Figure 9.1 from [3] shows a global vulnerability management process detailing the responsibilities of various organizations and services involved in ensuring system functionality.

- P1—Control and protection system engineer from the supplier/vendor.
- P2—Third-party service provider.
- P3—Control and protection system integrator.
- P4—Control and protection engineer from the energy company (EPU).

Vendors must have a clearly defined cybersecurity threat assessment program to promptly address vulnerabilities. This process requires up-to-date cybersecurity information from various third-party component providers (e.g., Microsoft, Sybase) and organizations such as CERT (Computer Emergency Response Team).

Collaboration Efforts

Effective cybersecurity solutions require collaborative efforts between control and protection engineers, network engineers, and other specialists with relevant expertise

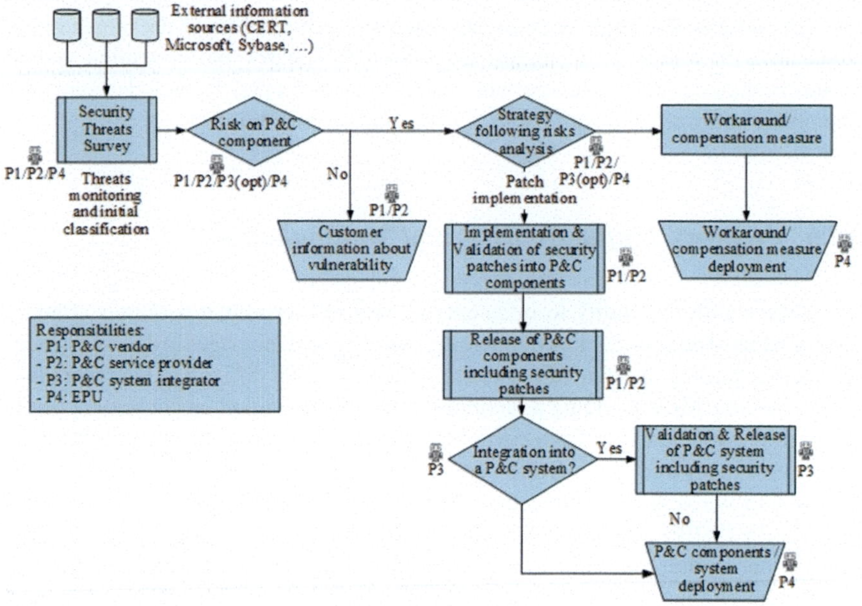

Fig. 9.1 Vulnerability management process

and skills. For example, managing network devices (routers, switches, firewalls) typically falls under the responsibilities of the network engineer. However, the control and protection engineer should work closely with the network engineer to ensure network device configurations do not compromise the performance, reliability, or availability of control and protection systems.

Physical Security of Control and Protection Systems

Physical security encompasses a range of measures, from locked gates and doors for accessing IEDs to preventing unauthorized access to automation systems. Access security to IEDs is achieved using role-based access control (RBAC), including password levels and user permissions. Utilities should maintain a database of personnel's permitted functions, associated with each individual's password. (Further details on RBAC aspects can be found in CIGRE Technical Brochure 427" Impact of Implementing Cybersecurity Requirements Using IEC 61850").

Common Malware Issues

Advanced malware spreads particularly effectively on endpoints that connect protection and control networks to unsecured devices. The independent research organization Ponemon Institute reports that their studies reveal frequent network incidents such as malware attacks, botnet attacks, and SQL injections. The most complex incidents involve zero-day attacks, SQL injections, and exploiting software vulnerabilities older than three months. Employees working remotely and downloading unfamiliar third-party applications pose a major concern, increasing the risk of destructive, hard-to-detect malware attacks. According to the 2010 Verizon Data Breach Investigations Report, common attack vectors include web applications, remote access services, remote management software, and backdoor or control channels.

A New Endpoint Security Stack

A new approach to addressing this issue is the creation of an endpoint security stack, which places patch and configuration management at its core, surrounded by layers of coordinated application, device, and antivirus management. Control and protection engineers should consider the following strategies and tactics:

- Enhanced Patch Management: IEC 62443-2-3 offers an improved approach to patch management, which should be considered.
- Antivirus Installation.
- Whitelisting Applications: Only approved applications can be executed, blocking everything else; by default, this protects the network, including from zero-day attacks, as it eliminates the need to wait for the latest vulnerability patch or antivirus definition.

Engineers should be proficient in patch management and perimeter antivirus protection and should understand whitelisting management procedures.

Network Security of Control and Protection Systems

Network security is an essential building block for ensuring the multi-layered protection of automation systems. Additionally, it plays a critical role in the conceptual design, as shown in Fig. 9.2 [4], influencing communication aspects. Major use cases include substation automation, substation-to-substation, substation-to-control centre, and remote engineering. In general, network security fulfils the following security functions:

- Access Control—reliable identification mechanisms for all elements of control and protection systems connected to networks: users, devices, and applications.
- Data Confidentiality and Integrity—encryption is not mandatory unless confidentiality is required by standards and regulations.

Operational Constraints: Ensuring Reliable Access to Substation Automation Resources

There is a clear need for comprehensive role-based access control management within substation automation systems. This should include various independent organizations essential to substation operations, such as support contractors and regulatory agencies that require legitimate access to substation assets and control. Some situations require on-site access, while others necessitate remote access. In any case, the control and protection engineer must declare individuals needing access to control

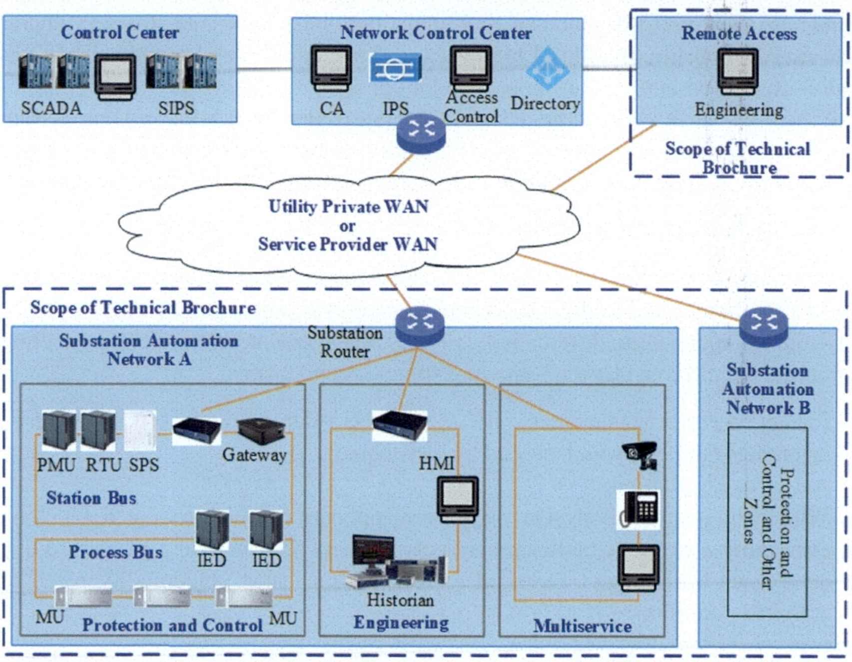

Fig. 9.2 Conceptual network security architecture

and protection equipment as untrusted. To ensure reliable access and utilization of substation assets, digital certificates should be employed to verify and manage user access rights and privileges. (Detailed information on lifecycle access control mechanisms for key management can be found in CIGRE Technical Brochure 603, "Application and Administration of Cybersecurity Measures for Protection and Control Systems", Appendix M).

Maximizing Compensating Security Measures

Control and protection engineers recognize three reasons for the need to implement compensating security measures for critical assets and functions.

Firstly, legacy systems, subsystems, and components lack adequate security mechanisms and must rely on perimeter protection systems.

Secondly, many new protection and control components lack the memory or processing resources to implement security mechanisms.

Thirdly, the response time for power system protection cannot afford communication delays or processing time required for tasks such as encryption and decryption to protect data confidentiality and integrity. For these reasons, control and protection engineers should emphasize maximum utilization of the first line of defence, specifically access control to the substation LAN through interfaces that are not considered sufficiently reliable. The second line of defence consists of multiple security mechanisms, providing usage control (e.g., read/write privileges specified in digital certificates), limited data flow managed by substation network routers, and network resource management embedded in all IEDs within the control and protection system.

9.3.2 Cyberattack Countermeasures by Substation Automation Engineers

Key tasks for specialists to address this issue include:

- Control and protection engineers (and IT engineers) should review and approve factory and on-site acceptance testing plans and procedures, including test scenarios, to ensure they incorporate comprehensive cybersecurity requirements.
- Although antivirus is ineffective against zero-day attacks, control and protection engineers must ensure that up-to-date patches are applied to perimeter protection systems to block known threats.
- Control and protection engineers should use security scanners to objectively assess the state of cybersecurity. They should periodically conduct selective scanning of individual sites and critical applications, perform automated penetration testing, and conduct system audits to support policies, procedures, and organizational measures required by regulations.
- In addition to perimeter antivirus protection, engineers should implement strict whitelisting policies to secure network access and protect control and protection components.
- Engineers should periodically review all wireless remote access to the control and protection system. Wireless access using Wired Equivalent Privacy (WEP) encryption should be considered outdated, and its interface with the system should be declared "untrusted".
- Role-based access control requires effective key management to protect access control and user management privileges specified in digital certificates issued to users. Key management is not solely the responsibility of control and protection engineers. However, these engineers should be part of the key management team and bear responsibility for influencing the design and operation of key management support functions directly related to substation automation systems, subsystems, and components that use key material for security protection.

CIGRE Technical Brochure 603, "Application and Administration of Cybersecurity Measures for Protection and Control Systems", describes ten essential considerations:

1. The scope and level of protection should be specific and relevant to the protection and control of threatened assets. The choice of means should be customized to each specific case and provide an appropriate context responsive to organizational structures and policy.

2. Cybersecurity mechanisms should be comprehensive, simple, scalable, and easily manageable by control and protection engineers within their routine duties.
3. Where applicable, based on the risk assessment of the power supply enterprise, intelligent electronic devices and applications should exchange data using open, secure protocols, such as those described in IEC 62351.
4. All protection and control devices should be capable of supporting their own cybersecurity policy (or providing compensating security) in an untrusted network.
5. All control and protection engineers, technical personnel, and managers, as well as their resources (including cybersecurity technologies), should have declared and transparent trust levels for any data exchange.
6. IEDs should provide appropriate levels of mutual authentication for access to systems and data.
7. Outside of the protection and control-specific domain, authentication, authorization, and accountability require close attention to the reliability of external interfaces.
8. In accordance with IEC 62351 attributes, access to protection and control data is controlled.
9. Data confidentiality (and cybersecurity of any high-value asset) requires segregation of duties and privileges, supported by strong role-based access control (RBAC) mechanisms.
10. Protection and control system data can only be protected when security mechanisms are enabled by default for data storage, transmission, or use.

9.3.3 Cybersecurity Testing Platform for IEC 61850 Digital Substations

Although IEC 62351 provides a framework for developing cybersecurity protocols for IEC 61850, challenges remain, and major manufacturers often fail to ensure adequate security in their intelligent electronic devices (IEDs). In recent years, during the construction of digital substations, manufacturers and power providers have focused more on device interoperability and function implementation than on addressing and testing cybersecurity. Nonetheless, cost-effective cybersecurity research for IEC 61850-based digital substations is still in its early stages. Much more in-depth research and analysis of specific vulnerabilities and cyberattacks are needed. To this end, this section presents a universal cyber-physical testbed for SCADA, designed to explore potential vulnerabilities through simulated cyberattacks. This testing environment allows for the analysis of real attack scenarios, as well as the proposal and evaluation of effective cybersecurity countermeasures for digital substations.

To investigate potential cybersecurity vulnerabilities in IEC 61850-based digital substations, the authors [5] constructed a cyber-physical testbed at the Jiangsu State

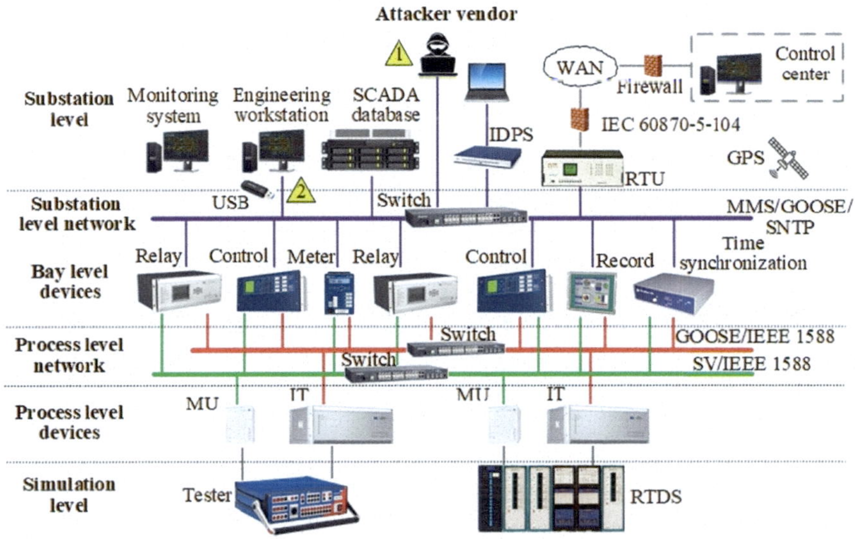

Fig. 9.3 Cyber-physical testbed for an IEC 61850-based intelligent substation

Electrical Power Research Institute (Nanjing, China), as shown in Fig. 9.3. The testbed comprises four levels: modelling, process, bay, and substation.

At the modelling level, a real-time digital simulator (RTDS) is used to simulate multiple power system event scenarios and model the transient characteristics and behaviour of simulated power systems. A universal relay testing and commissioning tool, serving as a programmable source of voltage and current, acts as an additional simulator for implementing stable and transient simulations.

At the process level, merging units (MU) are connected to the sampled value (SV)/IEEE1588 network, and intelligent terminals (IT) are connected to the Generic Object-Oriented Substation Event (GOOSE)/IEEE1588 network. The process-level networks are fibre-optic networks based on Ethernet switches. The bay-level IEDs include relays, measurement and control devices, a fault recorder, a network analyser, and a time synchronization IED. Bay-level IEDs connect to both process-level and substation-level networks.

The substation level includes a monitoring system, engineering workstation, SCADA database, remote terminal unit (RTU), and a laptop for launching cyber-attacks (the attack may propagate from a compromised node in the network, rather than directly from the laptop). The substation-level network, a cable network based on a switch, supports the Manufacturing Message Specification (MMS), GOOSE, and Simple Network Time Protocol (SNTP). The control center communicates with the intelligent substation via IEC 60870-5-104.

9.3.4 Cyber Vulnerability Studies on Digital Substations

The following types of cyberattacks are modelled and studied on the testbed.

Reconnaissance Attack. This attack allows an adversary to identify potential targets before an attack. In this testbed, a reconnaissance attack is launched using a laptop, indicated by the yellow triangle labelled "1" in Fig. 10.3, to gather online information about IEDs (IP addresses). For example, Havex malware is known to conduct reconnaissance. An adversary can use this information to launch more effective attacks.

Malformed Packet Attack. In this testbed, malicious IEC 61850 client software generates malformed packets based on the IEC 61850 protocol and sends them to intelligent electronic devices. Improper packet formats can disrupt the IEC 61850 protocol stack and cause IED communication failures or status exceptions, compromising the safe and reliable operation of the digital substation.

Denial of Service (DoS) Attack. A DoS attack is launched to capture all report control blocks of IEDs, which are created during configuration in the logical node (LN) on this testbed. Consequently, the targeted IED cannot respond to normal connection requests.

Address Resolution Protocol (ARP) Spoofing Attack. In the test environment, a laptop initiates an ARP spoofing attack by broadcasting ARP packets with the IP address of the monitoring system on the substation-level network. After the attack, IEDs interact with the malicious laptop instead of the monitoring system.

Man-in-the-Middle (MITM) Attack. This attack allows an adversary to intercept communication traffic between the monitoring system and IEDs, redirecting it to a malicious laptop on the substation-level network. The adversary can send malicious remote control commands and alter protective settings, impersonating the monitoring system; simultaneously, the adversary impersonates an IED and sends false, abnormal, or even distorted messages to the original monitoring system. In this test environment, MITM attacks can lead to abnormal behavior in the monitoring system and custom electronic devices and can even cause network failures.

Configuration Tampering. Substation Configuration Language (SCL) files form the basis for secure, stable, and reliable operation of IEC 61850-based intelligent substations. In this testbed, an adversary forges the configured IED description (CID) file in a protection relay; as a result, the relay malfunctions when simulated network failures occur.

Operating System/Database Attack. The operating system/database is exploited using known vulnerabilities on this testbed. For instance, VxWorks, a real-time embedded operating system in some IEDs, is targeted using the Wind River Debug vulnerability (WDV).

The potential consequences of cyberattacks on a digital substation, as described in this section, are shown in Table 9.1 [6].

According to the previously mentioned study on the testbed, potential attack scenarios on the substation are defined as follows:

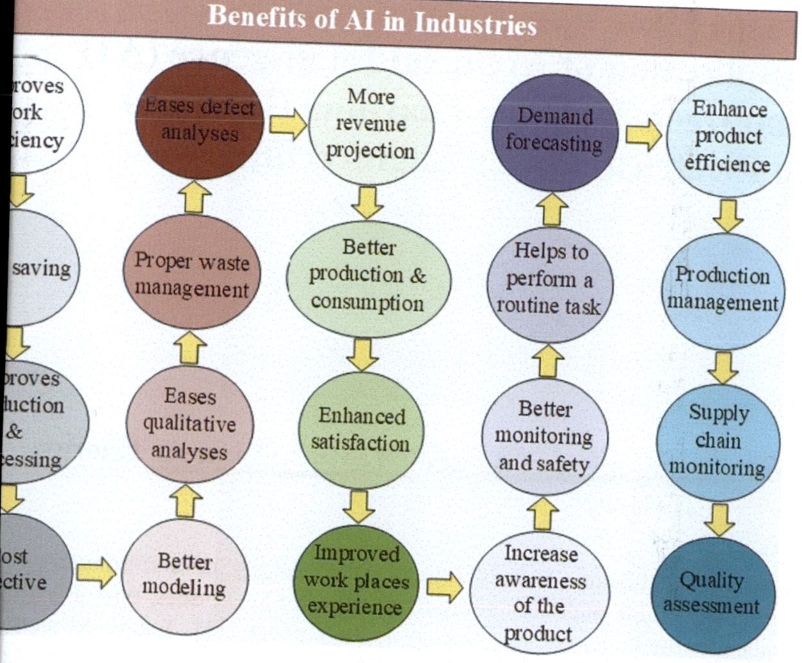

Benefits of AI in industries

ks to the proposed innovative solutions and the resulting increases in produc-
fficiency, and product quality, AI technologies are being implemented across
us sectors—from healthcare to finance. Of particular importance is the appli-
f AI to improve energy and resource efficiency in industrial enterprises.
ng to the definition by Islam & Hasanuzzaman, energy efficiency refers to
tion of the total energy input to a machine or system that is consumed in useful
d not wasted as useless heat or otherwise." [7] In accordance with this defi-
he goal of enhancing energy efficiency lies in reducing energy and resource
ption while maintaining the same level of product quality and technological
performance.

context of energy efficiency, AI—including its branch of Machine Learning
enables enterprises to implement the following key measures:

yse large volumes of data to identify patterns and trends. This helps opti-
supply chains, forecast demand, and detect hidden factors affecting both
uction efficiency and the efficient use of electricity and energy resources.
ct potential issues and prevent equipment downtime through predictive main-
ce. ML algorithms can analyse operational data from equipment and predict
ble failures before they occur, allowing maintenance and repairs to be planned
minimal disruption to the production process.

Table 9.1 Potential consequences of cyberattacks on digital substations

	MITM attack	Configuration tampering	Operation system/ database attack	Malformed packet attack	DoS attack	ARP spoofing attack
Failure to tripping protection	+	+				
Unwanted operation of protection	+	+				
Blocking protection	+	+				
Disruption of communication for protection	+	+	+	+	+	+
Network disruption within substation			+	+	+	
Abnormal Behavior or disruption of monitoring system	+	+	+	+	+	+
Denial of service from control center	+	+	+	+	+	+
Misjudgment of dispatchers	+					
Erroneous post analysis	+		+			

- Malware can spread through infected USB flash drives and local network connections, as shown by the yellow triangle labelled "2" in Fig. 9.3, and may be intended to sabotage SCADA systems with direct physical consequences.
- In this use case, a maintenance engineer's laptop is directly connected to the substation network switch, as shown in Fig. 10.3 (yellow triangle labelled "1").
- Due to unintentional misuse by the maintenance engineer, the laptop may gain access to unauthorized intelligent electronic devices (IEDs) from other manufacturers, potentially disrupting communications or even causing a system failure. Traditional malware, such as worms, or advanced malware similar to Stuxnet could infiltrate the IT environment of the digital substation via the laptop. Moreover, the same laptop could infect multiple digital substations maintained by the same engineer.

- During maintenance activities, the engineer may connect to the Internet via a wireless network to download electronic resources or perform remote maintenance. An adversary could exploit this as a channel to infiltrate the substation's automation system.

Digital substations are also vulnerable to further exploitation due to a lack of real-time detection methods for ensuring cybersecurity. Substation IEDs cannot determine whether control commands come from an authorized user or an "outsider", and IEDs in substations are typically not subjected to cybersecurity penetration testing before commissioning.

Control Questions and Tasks

1. What are the prerequisites for successful cyber protection?
2. Name the four categories of cyberattacks according to the CIGRE classification (CIGRE Technical Brochure 603, "Application and Administration of Cybersecurity Measures for Protection and Control Systems").
3. What actions should control and protection system engineers take to counteract cyberattacks?
4. How can cybersecurity be practically ensured for the control and protection system of a power facility (using the testbed example in Sect. 10.3.3)?
5. What types of cyber vulnerabilities are inherent to digital substations?

References

1. Trepanier T (2017) Digitalization: moving from step change to transformation in power generation. Powergen Int 121(7)
2. Ushakov V (2018) Electrical power engineering: current state. Springer, Problems and Perspectives, p 258
3. Gaspar J, Cruz T, Lam C, Simões P (2023) Smart substation communications and cybersecurity: a comprehensive survey. IEEE Commun Surv Tutor 25(4):2456–2493
4. Hokama WS, de Souza JS (2020) Cybersecurity for smart substation. IEEE PES Transm Distrib Conf Exhib 2020:1–6
5. Yang Y, Xu H, McLaughlin K, Sezer S, et al (2019) Cybersecurity testing technology in smart substations. IEC 61850-based smart substations. Academic Press, pp 223–254
6. Holik F, Yayilgan S, Olsborg G (2024) Emulation of digital substations communication for cyber security awareness. Electronics 13(12):2318

Chapter 10
Application of Artificial Intellige
and Deep/Machine Learning (DI
in Electrical Power Engineering

10.1 General Use of AI and DL/ML to Im
Efficiency

Artificial Intelligence (AI) is defined as the ability
machine to think and learn similarly, to how a human
embedded Artificial Neural Network (ANN) system [
which primarily encompass the functions of learning, e
decision-making—are extremely popular in various appl
to complement and enhance human cognitive abilitie
the Fourth and Fifth Industrial Revolutions (Industry 4
are becoming increasingly widespread and are signifi
industry, and technical systems. These technologies
ciency, the development of smarter products, and a r
[3].

The advancements in AI are closely linked to ne
processing large volumes of data, as well as progress in
and software development. These advancements enabl
with complex architectures, higher computational pow
algorithms. For both current and future benefits, indiv
industry giants (such as Google, Microsoft, Meta, IBM
investing heavily in AI development. Corporate investm
high of $189 billion in 2023, marking a 1300% increase
figures are expected to continue growing rapidly [4]. At
is observed across major industries: the highest profit
companies that invest more heavily in AI development.
for this trend is the fact that AI contributes to the cre
products, expands existing product lines, and significan
for companies [5]. The most evident advantages of u
industrial sector are shown in Fig. 10.1 [6].

V. Y. Ushakov et al., *Digitalization of Electrical Power Engineer*
https://doi.org/10.1007/978-3-031-95705-5_10

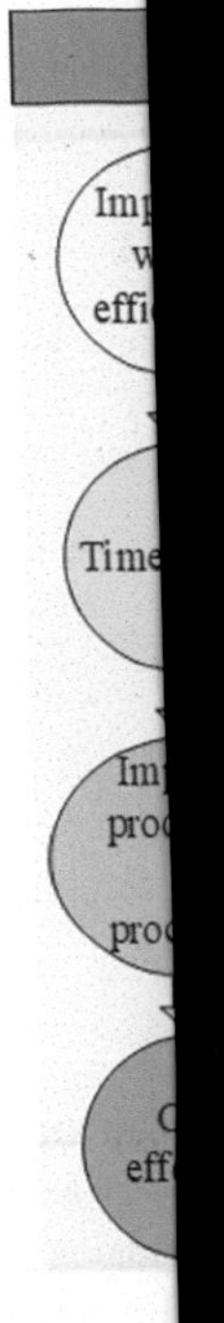

Fig. 10.1

Thar
tivity, e
numero
cation c
Accordi
"*the por*
work an
nition, t
consum
process
In th
(ML)—

- Anal
mize
prod
- Pred
tenan
possi
with

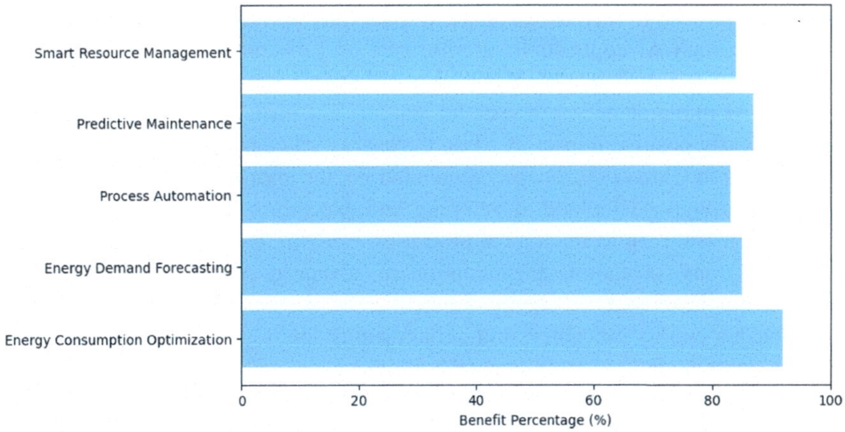

Fig. 10.2 Impact of AI and ML on electricity savings

- Optimize production processes by automatically adjusting parameters to achieve maximum efficiency. For example, AI can manage equipment settings to minimize energy consumption while maintaining the required level of performance.
- Enhance quality control using computer vision and deep learning to detect product defects with a level of precision exceeding human capabilities.
- Optimize logistics and production planning by considering multiple factors such as raw material availability, equipment load, product demand, and even weather conditions.

Figure 10.2 illustrates how AI and ML contribute to energy efficiency improvement in industrial enterprises through the implementation of five key measures.

Optimization of energy consumption through AI enables enterprises to reduce unjustified electricity usage and, consequently, lower energy-related financial expenditures. This is achieved through intelligent algorithms that manage energy consumption in real time, achieving an effectiveness of up to 92%.

Energy demand forecasting: Machine Learning (ML) facilitates the development of accurate models to estimate justified energy consumption and reduce excessive usage (effectiveness—88%) [8].

Process automation through AI allows enterprises to enhance production processes, making them more efficient and less energy-intensive (effectiveness—87%).

Predictive maintenance helps to forecast and prevent equipment failures, thus minimizing downtime and reducing repair costs (effectiveness—90%).

Smart resource management using AI enables companies to achieve a positive effect in 87% of cases.

Therefore, the implementation of AI technologies can significantly enhance the energy efficiency of industrial enterprises by optimizing processes and improving resource management. While this requires investments in technology and personnel training, it can lead to increased competitiveness and profitability in the long term. Importantly, such AI applications are not part of a distant future but represent a contemporary reality. There are comprehensive studies demonstrating the positive effects of AI implementation in specific regions. For instance, empirical evidence shows a clearly positive impact of AI on the energy efficiency of enterprises across 30 provinces in China [9]. Key recommendations for improving enterprise energy efficiency through AI include: developing and expanding state-supported energy efficiency measures in technological processes, bridging the growing gap between different regions, promoting the adoption of advanced AI technologies in less economically developed and energy-deficient areas, accounting for adjacent factors that influence energy efficiency, and strengthening interregional and international cooperation in the field of intelligent technologies.

10.2 Need for AI and DL/ML in Smart Electric Grid

Artificial Intelligence (AI), as a multifunctional technology, is becoming increasingly widespread across various industrial applications and economic sectors. The electrical power sector is no exception, as the application of AI has become fundamentally important due to the rapidly rising demands on increasingly complex power systems. The challenges faced by modern power systems stem from the integration of new, including intelligent, equipment, the use of renewable energy sources and energy storage systems, the growing influence of consumers on grid management, the need for processing and exchanging large volumes of data, the adoption of new modelling and computational methods, and the transformation of fundamental grid operation approaches [10]. In this context, the Smart Grid concept stands out as an innovative power system architecture capable of automatically adapting to changing conditions and optimizing resource usage. In addressing all of the above challenges, the role of AI technologies becomes central.

In the power sector, a wide array of intelligent technologies can be applied, including Artificial Intelligence (AI), Deep Learning (DL), Machine Learning (ML), metaheuristic optimization algorithms, and other AI methods such as fuzzy logic, reinforcement learning, and model predictive control [11]. The integration of such advanced approaches offers significant advantages over traditional technologies. AI, in particular, is capable of operating within modern, complex, and large-scale energy systems, especially in relation to the associated big data. AI can efficiently process such information, identifying patterns across numerous interconnected processes. Beyond analytics, AI shows potential as a real-time decision-making tool. Traditional control and regulation methods rely on pre-defined rules and parameters, which may be insufficient in certain critical scenarios. AI algorithms demonstrate adaptability and more advanced responsiveness under dynamic conditions, thanks to their ability to learn from new data and incorporate prior experience. The predictive function is another major advantage of AI, allowing for the construction of accurate forecasts of future conditions and operating modes, which facilitates the prevention of

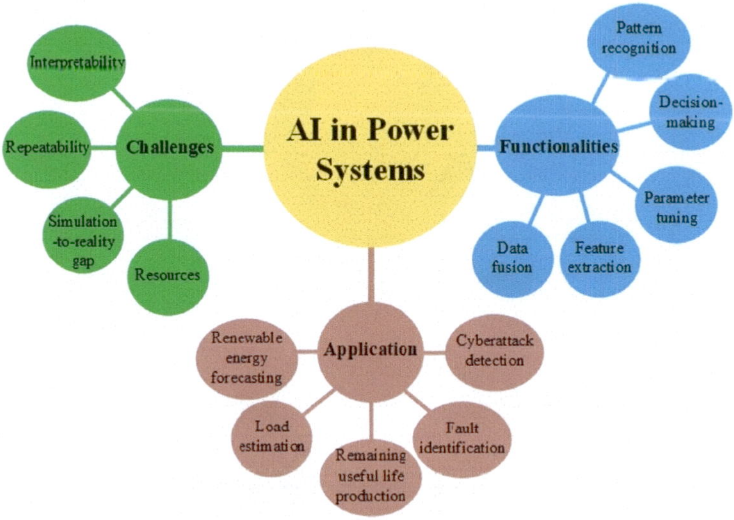

Fig. 10.3 AI in electrical power engineering

failures. Finally, AI excels at managing uncertainty, i.e., handling fluctuating and unpredictable parameters—particularly relevant in power systems due to variable loads and the intermittent nature of renewable energy sources [11]. Today, the capabilities, limitations, and functional roles of AI are the subject of intensive study in numerous scientific investigations. An overview of AI applications in electrical power engineering is clearly represented in the schematic diagram in Fig. 10.3 [12].

Figure 10.4 presents an overview of the potential applications and benefits of various AI technologies in the electrical power engineering. First, due to its predictive capability, AI can perform fault detection, i.e., identifying and forecasting potential failures of critical power equipment before an actual fault or outage occurs. This is made possible through the comprehensive analysis of current and historical big data from sensors, allowing detection of patterns in equipment condition, anomaly identification, and early recognition of emerging issues. Second, based on AI-powered forecasts, predictive maintenance can be performed, significantly increasing equipment reliability and overall grid resilience. Third, AI technologies are capable of optimizing the processes of electricity generation, consumption, and distribution, as they can account for numerous real-time variables in load forecasting, including demand trends, weather changes, and the status of generating assets, including renewable energy sources. Fourth, energy flow management within the grid becomes more sophisticated with AI, allowing for the minimization of energy losses through rational distribution and the most efficient use of available resources. Fifth, AI proves to be a powerful tool in addressing cybersecurity challenges, as it can analyse diverse data streams to detect suspicious activities that may indicate potential cyberattacks.

All of these AI applications can significantly enhance the reliability and security of power systems, including reducing equipment downtime and maintenance costs,

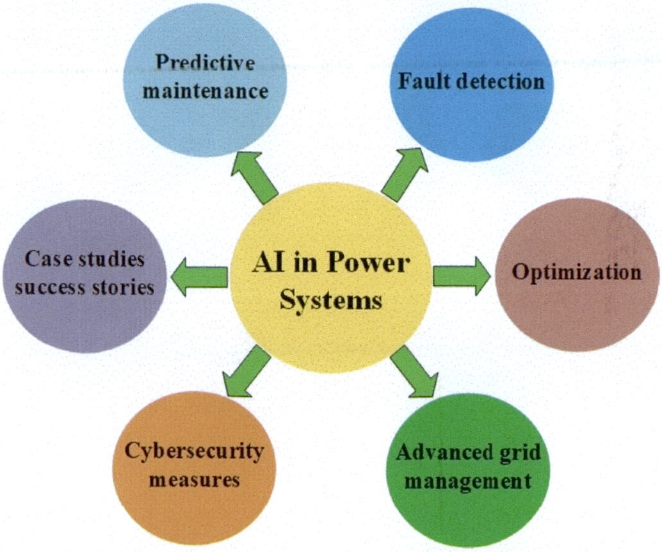

Fig. 10.4 Potential applications of AI in electrical power systems

while improving energy and resource efficiency. This fully aligns with the Smart Grid paradigm, which involves monitoring, and measuring real-time processes, enabling data transmission to control centres, automatic response actions, communication between devices and systems, and processing and analysis of data from digital sensors [13]. Thus, AI technologies are central to the full implementation of Smart Grid concepts, offering effective tools for data processing, analysis, and control, as well as real-time decision-making based on this information.

10.3 Types of AI and DL/ML Techniques for Power System

Artificial Intelligence (AI) encompasses a broad set of methods that enable computer systems to perform various cognitive tasks by simulating human behaviour and intelligence. This offers a unique opportunity to solve applied problems across multiple domains of human activity, including the electrical power sector. Often, the term Machine Learning (ML) is used synonymously with AI. However, in reality, ML is a subset of AI that involves techniques enabling computers to learn and improve from large datasets without explicit programming [14]. Deep Learning (DL) algorithms, as a specialized form of ML, are distinguished by their use of multi-layered interconnected neural networks to extract patterns from massive amounts of unstructured data, encompassing both input and output layers. In Fig. 10.5 provides a conceptual illustration of the terms AI, ML, and DL [15]. Thus, ML and DL are both considered subdisciplines within the broader field of AI.

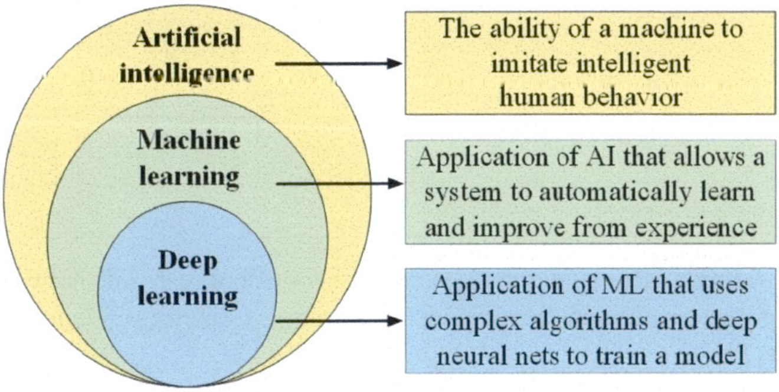

Fig. 10.5 Concepts of AI, DL, and ML

In the context of their application to power systems, the use of Artificial Intelligence (AI) encompasses four main categories of techniques: Artificial Neural Networks (ANNs), Fuzzy Logic, Expert Systems, and Evolutionary Methods. These techniques, whether applied individually or in combination, can significantly enhance system reliability and stability.

Artificial Neural Networks (ANNs) are mathematical models or algorithms inspired by the structure and functioning of neural networks in the human brain (Fig. 10.6) [15]. They consist of a large number of interconnected simple processing units called neurons. Each fundamental neuron receives input data, processes it using a non-linear transformation, and produces an output signal.

These output signals are then transmitted as inputs to other neurons. This process continues until the network produces a final result. ANNs can vary in structure, processing approach, direction of information flow, and the number of layers. One of the most important features of ANNs is their ability to learn from large datasets, recognize patterns, and make predictions. This makes them highly effective in tasks such as pattern recognition, natural language processing, and forecasting. Studies have shown that ANN functions can be successfully applied to power system tasks

Fig. 10.6 Typical structure of ANN

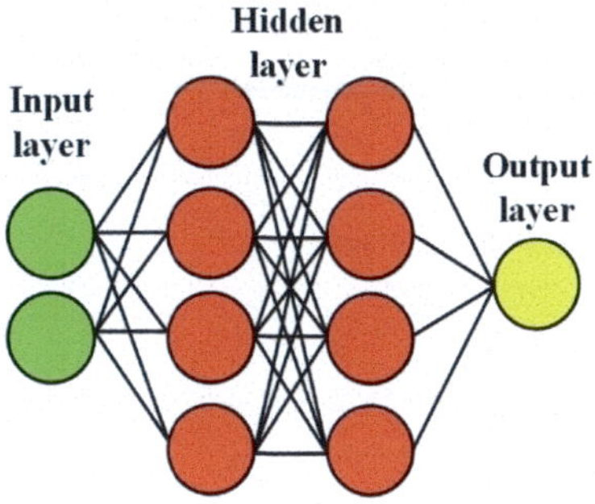

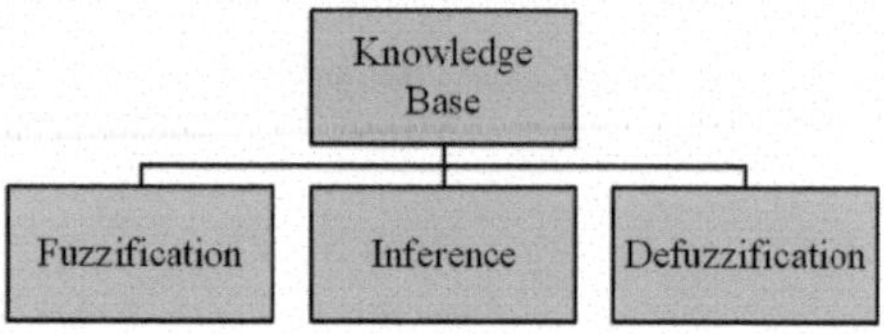

Fig. 10.7 Fuzzy logic system

such as operating mode calculations, intelligent control of real-time network parameters and power equipment, and modelling complex interdependencies between components [15].

AI techniques based on *Fuzzy Logic* (FL) are designed to describe and model systems characterized by uncertainty and imprecision (Fig. 10.7) [16]. These techniques enable the standardization of approximate reasoning and the formulation of precise conclusions in a manner that resembles human decision-making strategies. As a result, complex real-world physical and technical systems—which often involve ambiguity and operate with fuzzy data—can be effectively analysed. The electric power industry is a prime example of such a system. In this domain, FL techniques can be effectively applied to control processes, where it is necessary to consider numerous factors and make decisions based on incomplete or imprecise information. In addition, fuzzy logic is used for the processing and transformation of numerical and symbolic data [15].

In *Expert Systems* (ES), knowledge bases and logical rules are used to solve specific domain problems automatically (Fig. 10.8) [13]. These systems emulate the knowledge and experience of a human expert, providing recommendations based on available data. They process large volumes of information, detect patterns, and generate conclusions. Typically, the use of expert systems is essential for solving tasks that exceed the capabilities of human intelligence and experience. As such, expert systems can be used for real-time decision-making, knowledge archiving, and solving a wide range of problems that require both logical reasoning, heuristics, and inference, as well as fast processing of large-scale data. In the context of electrical power engineering, expert systems are used for decision-making based on calculated and forecasted operational parameters of the system during processes such as electricity generation, transmission, and distribution.

Evolutionary AI methods are based on the fundamental principles of natural evolution: heritability, variability, and natural selection (Fig. 10.9) [17]. These principles

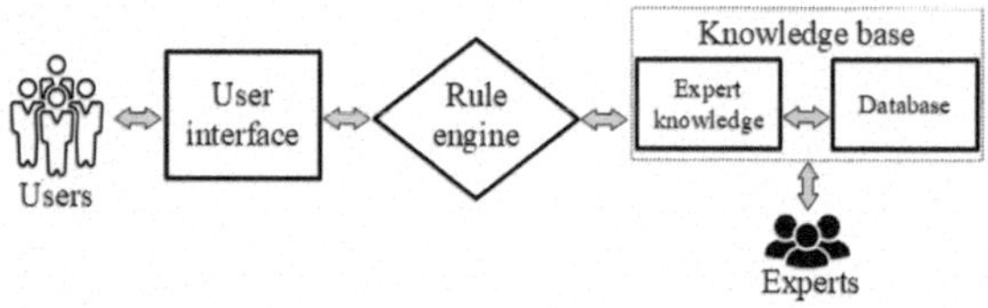

Fig. 10.8 ES diagram

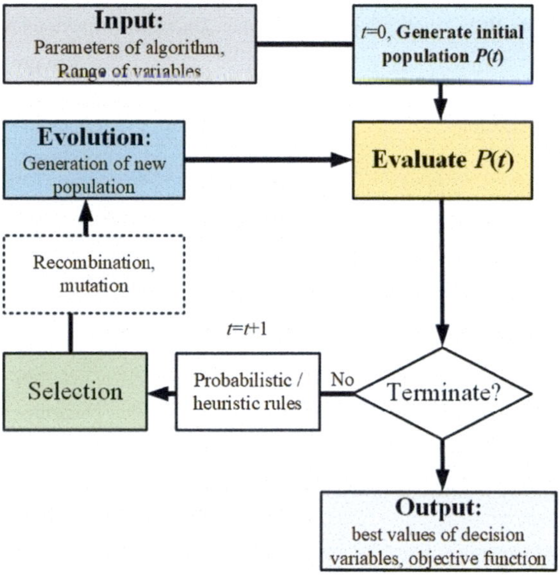

Fig. 10.9 The basic structure of an evolutionary algorithm (EA)

make evolutionary methods a powerful approach for optimizing various domain-specific systems with minimal computational effort and straightforward implementation. Among the primary evolutionary techniques are: Genetic Algorithms (GAs), Evolutionary Programming (EP), and Differential Evolution (DE). These methods can be applied to a wide range of problems without requiring deep mathematical expertise, which is particularly important when solving complex optimization tasks in power systems. The effectiveness of GAs has been demonstrated in numerous applications, such as placement of renewable energy systems and determination of their optimal operating modes, power flow analysis, reactive power compensation, maintenance scheduling for electrical equipment, load management, reduction of electrical energy losses and improvement of power supply quality indicators.

The AI techniques discussed above are by no means exhaustive in terms of their applicability to power system problems. Additional techniques that have also demonstrated effectiveness include the Minimum Spanning Tree (MST), Simulated Annealing, Particle Swarm Optimization (PSO) [8], Ant Colony System (ACS) algorithm, and many others. The implementation level of various AI techniques (as of 2022) is presented in Table 10.1 [18].

Table 10.1 Application and implementation level of various AI techniques

Power system supply chain element	Use cases	Most prominent AI domain	Research maturity
Generation	Improved planning	Planning	Low
	Improved O&M	Learning	High
Transmission networks	Improved planning	Planning	Medium
	Improved O&M	Learning	High
	TSO-DSO coordination	Planning	Low
	Grid automation	Learning	Low
Distribution networks	Improved planning	Planning	Medium
	Improved O&M	Learning	High
	Grid automation	Learning	Low
Isolated grids/ microgrids	Improved planning	Planning	Low
	Improved O&M	Learning	High
	Grid automation	Learning	Low
Market operations/ trading	Price forecasting	Learning	High
	Aggregation and flexibility	Learning	High
	Energy communities/ prosumers/local energy markets	Integration & interaction	Low
	Automated trading	Learning	Medium
Retail	Improved planning	Learning	Low
	Improved O&M	Learning	Low
	Customer/Demand/Load analysis	Learning	High

10.4 Role of AI and DL/ML for Power System Stability, Control, Security and Protection

The intensive discussion regarding the potential of AI technologies in the power sector began as early as the 1980s, when CIGRÉ established Working Group 38.06 titled "Expert System in Power System." With the advancement of computing hardware, increased processing power, and enhanced capabilities for handling large datasets, AI applications have gained new momentum and promise.

Various AI techniques are capable of solving a wide spectrum of problems related to the operation of electrical equipment, power system control, and planning. Figure 10.10 illustrates the potential applications of AI across different areas of electrical power engineering [16]. In practice, the continuous and reliable supply of high-quality electrical energy, along with efficient generation, optimal operational control, loss minimization, and affordability, is critical to the operation of any power system. In each of these areas, the potential role of AI is remarkably significant—especially under current conditions characterized by: increasing grid complexity, rising demand for electricity, the growing integration of renewable energy sources, currently, there is a strong and ongoing research focus on exploring the applicability of AI technologies in power systems. The most substantial emphasis is placed on AI applications in: ensuring power system stability, monitoring and controlling operational parameters, enhancing system and equipment security, including cybersecurity, protecting against faults and failures, and preventing emergency or abnormal operating conditions.

One of the principal tasks in managing an entire power system is to maintain the balance between generated and consumed power, taking into account all types of losses in the grid. If this balance is not ensured, the power system experiences deviations of operational parameters from nominal values (in particular, frequency and voltage). Traditionally, automatic systems based on linearized models and approximations of nonlinear processes have been used to control power generation. In this context, AI offers a higher level of accuracy in managing generation and consumption due to its advantages in optimization and solving complex nonlinear problems. Here, various ANNs and GAs have demonstrated the highest efficiency. Moreover, AI technologies can significantly assist in the integration of renewable energy installations into traditional power systems, as they can enhance the predictability of such

Domain	Operation	Control	Planning
	• Operations • Services • Customers • Distribution • Transmission • Generation	• Central computer • Terminal • Power station • Load • Digital reality • Relay control • Central control	• Generation • Transmission • Distribution • Customers
Application	• Economic load dispatch • Scheduling for generator maintenance • Power flow • Unit commitment	• Voltage control • Power system stability control • Load frequency control • Systems and services	• Planning for distribution • Capacitor placement • Optimization for reactive power • High penetration photovoltaics • Renewable generation methodologies
AI method	• Particle swarm optimization • Adaptive neuro-fuzzy interface system (ANFIS) • Mixed integer linear programming (MILP) • Genetic algorithm (GA)	• Hybrid networks with PSO • Artificial neural network (ANN) and PSO • Back propagation ANN • The neuro-fuzzy system with PSO • Nonlinear control	• PSO • Fuzzy logic and PSO • MILP with GA • ANN with GA • Ant colony • Optimization with GA

Fig. 10.10 Applications of AI in various areas of electrical power systems

sources. This is particularly relevant to wind generation, which depends on wind speed and direction—both of which can be accurately predicted using ANNs and supervised learning techniques.

In emergency situations, system restoration is typically carried out by automatic devices and relay protection. After a failure, the power system must be restored, i.e., returned to a regulated operating state regardless of the severity of the emergency. For many industrial sectors, delays in electricity supply can be critical; therefore, power restoration must occur immediately. This task is often complex and may require a revision of the current power distribution scheme, where evolutionary algorithms in combination with FL can be of assistance.

Assessing the current level of reliability of operating power systems is also a significant task that a human operator alone cannot handle. AI technologies are capable of evaluating system reliability and maintaining it within required thresholds, including through ensuring system stability. Stability is determined by the parameters, scale, and duration of disturbances. Their detection and prevention require continuous monitoring of system parameters and their slightest changes. ANNs, in particular, have been shown capable of assessing stability during transient conditions.

Maintaining the required voltage at various points in the power grid is a complex and multifactorial problem. AI technologies allow the simultaneous activation of multiple voltage regulation mechanisms to keep voltage within required ranges—primarily by jointly controlling reactive power levels via compensating devices (e.g., STATCOMs) and by adjusting transformer tap changers. FL technologies are especially suitable for this task due to the complexity and ambiguity of the problem.

In the domain of power system protection, AI technologies offer effective methods for accurate and ultra-fast fault detection and prevention of emergencies by processing and analysing vast volumes of both historical and real-time data. ANNs can perform monitoring, control, and decision-making functions. Additionally, with the help of AI (especially FL), it is possible to implement tasks such as: dynamic adjustment of relay protection settings in response to the system's continuously changing parameters (FL), coordination and selectivity of protection systems across different regions (GAs) and reduction of repair time by accurately identifying faults in transmission and distribution networks (expert systems).

The review [11] presents the key roles of AI in various aspects of power system stability, control, and protection that can be implemented using different technologies. AI offers far more than just the functions described above for power systems. Emerging tasks addressed by AI include reducing the carbon footprint of energy systems, electricity price forecasting and improving energy efficiency among household consumers (Fig. 10.11) [15]. Moreover, AI technologies can be applied to tasks across different time horizons—ranging from real-time decision-making to long-term forecasting and planning spanning years or even decades (Fig. 10.12) [12].

Additionally, it is important to consider the role of AI tools in training personnel of power companies to solve practical cases. Based on the experience and big data accumulated through AI, AI algorithms can be used to simulate real energy facilities and entire power systems, enabling the creation of digital twins. These digital models allow staff to improve decision-making quality and efficiency. Furthermore, Machine

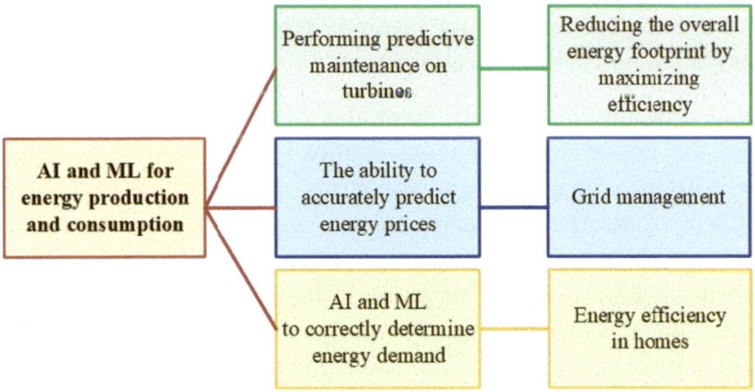

Fig. 10.11 AI and ML for energy production and consumption

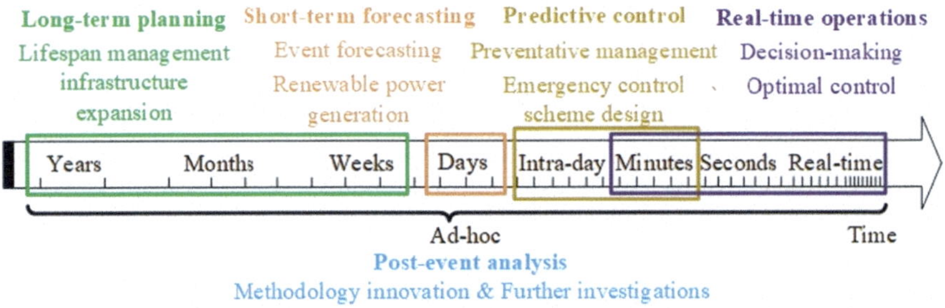

Fig. 10.12 Time-scale classification of artificial intelligence applications in power systems

Learning (ML) methods can utilize the results of such processes to continuously enhance the training process through data processing and analysis [11].

10.5 Today's Challenges and Future Scope of AI in Power System

The demonstrated capabilities of AI across various industrial sectors—and particularly in power systems—are driving its gradual adoption in real-world applications. Today, AI is already used in practice in: medicine—for disease diagnosis, economics—for market analysis, automotive industry—for autonomous driving, agriculture—for optimizing crop yields [19]. In the power sector, successful AI use cases include the following companies [11]:

- Siemens and General Electric (GE)—for more accurate electricity demand forecasting,
- Google DeepMind and the UK National Grid—for forecasting the amount of renewable energy generation based on current and predicted weather conditions,

– ABB and Schneider Electric—for planning preventive maintenance and repairs of power station and substation equipment.

At the same time, despite the high potential of AI technologies and these successful applications, the power sector remains relatively conservative in adopting new approaches. This is primarily due to the stringent requirements for the safe and reliable delivery of electricity to consumers, along with the associated risks. As a result, AI technologies must first be brought to an acceptable, near-perfect level of maturity, ensuring they can fully or partially replace traditional methods without compromising the current standards of control, safety, or reliability.

Therefore, a prerequisite for the full-scale adoption of AI in the power sector is the availability of compelling evidence of its effectiveness in specific areas—such as load and generation forecasting, control systems, and beyond. In addition, the successful functioning of AI technologies requires the creation, maintenance, and continuous updating of reliable and relevant datasets, as well as unhindered access to these datasets. Only under these conditions can AI fully unlock its potential, continuously self-learning and improving with new data. Moreover, inaccuracies may exist in the AI models themselves, requiring thorough refinement. Additional associated challenges include the development and support of high-quality software capable of implementing AI models, and the training and involvement of highly qualified personnel for developing AI technologies, building and managing datasets, and applying AI solutions in industrial settings [16]. Naturally, substantial investments are needed to address all of these challenges. These and other current issues associated with AI deployment are summarized in Fig. 10.13 [11].

However, all current challenges are, to varying degrees, solvable in the foreseeable future. The process of implementing AI in any field appears extremely promising due to the unique ability of such technologies to solve complex and ambiguous problems. For the power sector, the implementation of AI is one of the most critical areas of digitalization, driven by the rapidly increasing complexity of energy systems (especially in control systems), growing loads, expanding data volumes, the integration of new types of energy sources, and evolving relationships between the power system and the consumer. As a result, the introduction of AI may offer a historic opportunity to address many contemporary and long-standing problems in power engineering, which, in turn, is significant not only for the energy sector but also for related fields and overall economic development.

Control Questions and Tasks

1. What explains the extremely high interest in AI across various industrial sectors?
2. What is the value of using AI in power system applications (including in the context of the Smart Grid concept)?
3. List and characterize various types of AI that can be applied to solving power system problems.
4. How can the implementation of AI algorithms improve the stability, security, and reliability of power systems?

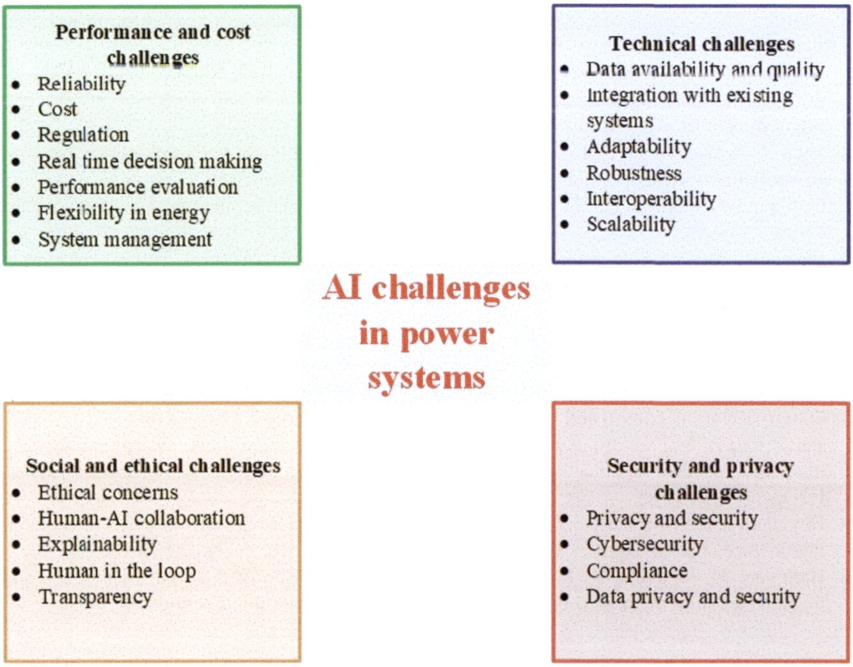

Fig. 10.13 Current AI challenges in power systems

5. What challenges are currently faced in the process of AI integration into power systems?

References

1. Zohuri B, Behgounia F (2023) Application of artificial intelligence driving nano-based drug delivery system. A handbook of artificial intelligence in drug delivery, pp 145–212
2. Li Y, Ding Y, He S, Hu F et al (2024) Artificial intelligence-based methods for renewable power system operation. Nat Rev Electr Eng 1:163–179
3. Leng J, Zhu X, Huang Z, Li X et al (2024) Unlocking the power of industrial artificial intelligence towards Industry 5.0: insights, pathways, and challenges. J Manuf Syst 73:349–363
4. Liu Z Leo (2025) Artificial intelligence for engineers
5. Babina T, Fedyk A, He A, Hodson J (2024) Artificial intelligence, firm growth, and product innovation. J Financ Econ 151:103745
6. Malik S, Muhammad K, Waheed Y (2024) Artificial intelligence and industrial applications-a revolution in modern industries. Ain Shams Eng J 15:102886
7. Islam MM, Hasanuzzaman M (2020) Introduction to energy and sustainable development. Energy for sustainable development: demand, supply, conversion and management, pp 1–18
8. Agostini L, Filippini R (2019) Organizational and managerial challenges in the path toward Industry 4.0. Eur J Innov Manag 22:406–421
9. Li X, Li S, Cao J, Spulbar AC (2025) Does artificial intelligence improve energy efficiency? Evidence from provincial data in China. Energy Econ 142:108149

10. Sharifzadeh A, Ameli MT, Azad S (2021) Power system challenges and issues. Power Syst 1–17
11. Alhamrouni I, Abdul Kahar NH, Salem M, Swadi M et al (2024) A comprehensive review on the role of artificial intelligence in power system stability, control, and protection: insights and future directions. Appl Sci 14:6214
12. Shen Z, Arraño-Vargas F, Konstantinou G (2023) Artificial intelligence and digital twins in power systems: trends, synergies and opportunities. Digital Twin 2:11
13. Omitaomu OA, Niu H (2021) Artificial intelligence techniques in smart grid: a survey. Smart Cities 4:548–568
14. França RP, Borges Monteiro AC, Arthur R, Iano Y (2021) An overview of deep learning in big data, image, and signal processing in the modern digital age. Trends Deep Learn Methodol Algorithms Appl Syst 63–87
15. Raihan A (2023) A comprehensive review of artificial intelligence and machine learning applications in energy consumption and production. J Technol Innov Energy 2:1–26
16. Pandey U, Pathak A, Kumar A, Mondal S (2023) Applications of artificial intelligence in power system operation, control and planning: a review. Clean Energy 7:1199–1218
17. Janga Reddy M, Nagesh Kumar D (2020) Evolutionary algorithms, swarm intelligence methods, and their applications in water resources engineering: a state-of-the-art review. H2Open J 3:135–188
18. Heymann F, Quest H, Lopez Garcia T, Ballif C et al (2024) Reviewing 40 years of artificial intelligence applied to power systems–a taxonomic perspective. Energy AI 15:100322
19. Hallmann M, Pietracho R, Komarnicki P (2024) Comparison of artificial intelligence and machine learning methods used in electric power system operation. Energies 17:2790

Chapter 11
Impact of the Application of Various Digitalization Tools on the Energy Efficiency of Industrial Enterprises

11.1 Internet of Things (IoT) for Resource Management

The IoT represents a network of interconnected devices capable of collecting and sharing data without human intervention [6]. In industrial enterprises, this includes:

- Smart sensors embedded in equipment for real-time monitoring of production processes and equipment status. These sensors can track parameters such as temperature, pressure, vibration, and energy consumption, enabling prompt responses to deviations from the norm;
- Automated control systems capable of optimizing processes based on received data. For example, a system may automatically adjust production line speed depending on the quality of incoming raw materials or current product demand;
- Enhanced asset and inventory tracking, leading to more efficient resource management. Radio Frequency Identification (RFID) and Global Positioning System (GPS) technologies enable real-time tracking of equipment, tools, and materials, minimizing losses and downtime;
- The creation of "smart factories", where all production elements are interconnected in a single network, providing an unprecedented level of control and optimization.

Figure 11.1, sourced from [7–9], illustrates how IoT technology contributes to increased energy efficiency in industrial enterprises through the implementation of five specific measures. (In Fig. 11.1 and subsequent figures in this chapter, the x-axis represents the values of actual electricity savings achieved through the implementation of specific measures as a percentage of the maximum possible savings).

1. ***Equipment Monitoring*** using IoT enables enterprises to track the condition of machinery and equipment in real time. This helps prevent breakdowns and reduces excessive energy consumption (effect—90%).

V. Y. Ushakov et al., *Digitalization of Electrical Power Engineering*, Power Systems,
https://doi.org/10.1007/978-3-031-95705-5_11

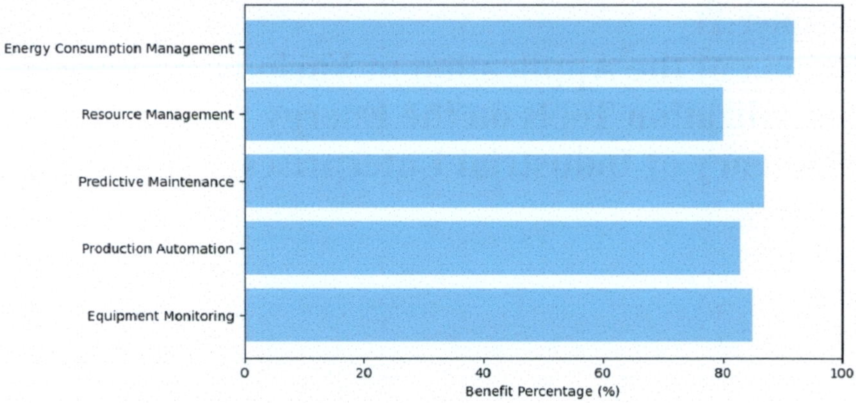

Fig. 11.1 Impact of the Internet of Things (IoT) on electricity savings

2. ***Production Automation*** with IoT technologies allows optimization of production processes, reducing energy consumption and lowering costs (effect—85%).
3. ***Predictive Maintenance***, enabled by IoT sensors and predictive systems, can foresee equipment failures. This prevents unscheduled downtimes and minimizes energy used for repair and maintenance, thus reducing excessive energy consumption (effect—88%).
4. ***Resource Management*** via IoT allows monitoring of energy and other resource usage in production processes. This contributes to a more rational distribution of energy costs and reduces unwarranted energy consumption (effect—80%).
5. ***Energy Consumption Management*** with IoT helps enterprises track and regulate energy use, reducing costs and enhancing overall energy efficiency (effect—92%).

11.2 Analysis of Energy Consumption Using Big Data

Using big data allows companies to:

- Deepen knowledge of production processes, which allows identifying inefficient stages, bottlenecks and opportunities for optimization.
- Make more informed decisions based on real data rather than intuition. This applies to both strategic decisions at the enterprise management level and tactical decisions at the production management level on the ground [10].
- Improve product quality by analyzing the factors that affect the final result. For example, it is possible to identify correlations between the quality of raw materials, production process parameters and the quality of finished products.
- Optimize energy consumption and resource efficiency by analyzing patterns in the use of energy and materials.

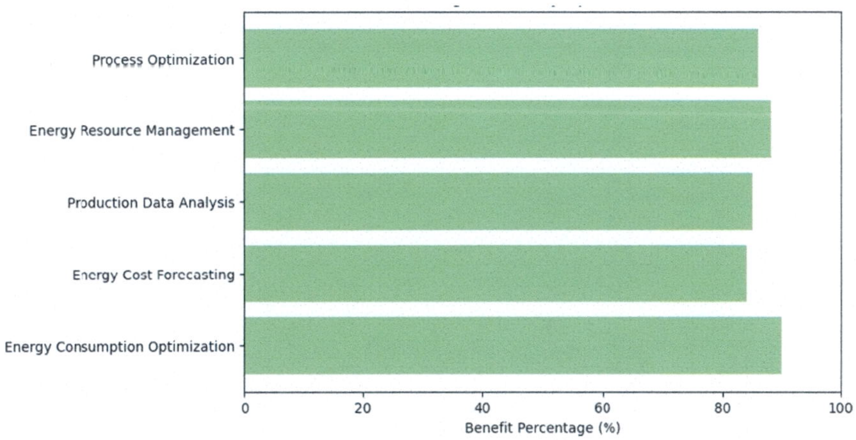

Fig. 11.2 Impact of big data and analytics on electricity savings

- Predict market trends and consumer behavior, which allows for more accurate production planning and updating the product line.

Figure 11.2 illustrates the impact of big data and analytics on improving energy efficiency in industrial enterprises through various key management decisions.

The effects of this impact are listed and briefly commented on below:

Energy Consumption Optimization using Big Data allows enterprises to more accurately manage energy consumption, which reduces costs and enhances overall efficiency (effect—89%).

Energy Cost Forecasting with Acceptable Accuracy becomes possible through analytical tools based on big data (effect—85%).

Production Data Analysis with Big Data reveals hidden patterns and factors affecting energy efficiency, helping enterprises improve technological and management processes (effect—87%).

Energy Resource Management based on Deep Analytics enables enterprises to effectively control and allocate energy resources (effect—90%).

Finally, **Process Optimization** using Big Data Analytics improves production processes, minimizes energy costs, and increases efficiency (effect—88%).

11.3 Cloud Technologies for Making Efficiency Gains

Cloud technologies provide the following benefits:

- Flexibility and Scalability of IT Infrastructure. Enterprises can quickly scale up or down their computing resources based on current needs without the need to invest in expensive hardware;

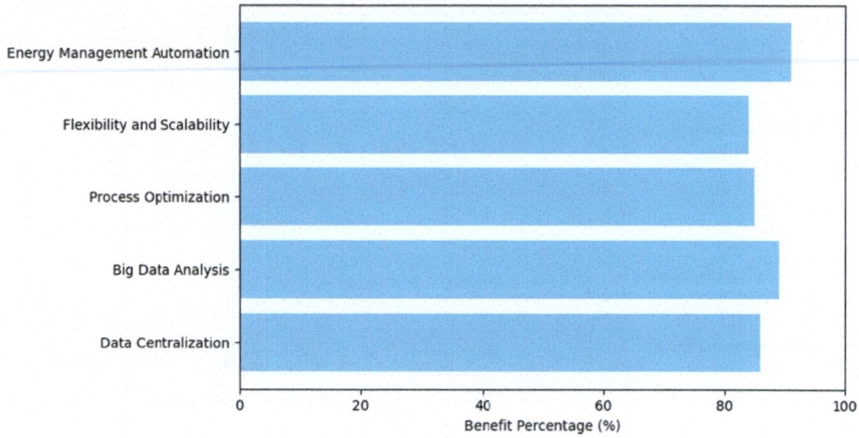

Fig. 11.3 Impact of Cloud Technologies on Electricity Savings

- Closer Collaboration and Data Sharing between different departments and even geographically distributed production sites. This is especially important for large corporations;
- Cost Reduction in IT due to the "pay-as-you-go" model. Enterprises can optimize expenses by paying only for the resources they actually use;
- Enhanced Data Security through advanced protection technologies provided by major cloud providers;
- Access to Advanced Technologies and Services, such as AI and machine learning, without the need to develop them in-house.

Figure 11.3 illustrates the significant impact of cloud technologies on improving energy efficiency in industrial enterprises.

Firstly, cloud solutions enable centralized data management, simplifying the monitoring and control of energy consumption within an enterprise. This facilitates more efficient analysis and adjustment of processes, resulting in significant energy savings [11].

Cloud technologies also provide enterprises with access to powerful big data analysis tools, allowing them to uncover hidden inefficiencies and optimize energy costs. This approach enables more informed decision-making aimed at cost reduction and overall productivity improvement.

Optimization of production processes through cloud technologies promotes a rational allocation of resources, which reduces excess energy consumption and enhances production outcomes. Furthermore, cloud platforms offer flexibility and scalability, helping enterprises adapt to changes in production volume and product range, and efficiently manage energy consumption even under variable loads.

Digitalization of energy management allows for reduced excess energy expenses through improved control over its usage.

11.4 Models and Digital Twins of Power Systems and Industrial Processes

One of the most advanced methods for studying both physical and industrial processes is modeling. During the modeling process of an object or phenomenon, the original is replaced by a model—either abstract or material—that can simulate its properties and characteristics for solving specific theoretical or applied problems. Investigating such models enables researchers to draw conclusions about real-world processes, which becomes particularly relevant when examining complex technical systems where significant constraints exist regarding actual physical experimentation (such as electrophysical phenomena [12–16]). Electric power systems are especially limited in experimental possibilities due to the continuous production process involved in electricity generation, transmission, and consumption, the interconnectedness of all system components, the impossibility of isolating individual elements, and the severity of potential consequences during emergency situations. Thus, developing mathematical and other models that closely approximate reality becomes the optimal solution. Nevertheless, modeling power systems is inherently challenging, as it must account for numerous uncertainties related to various phenomena such as load fluctuations, equipment faults and failures, and changing weather conditions, which also impact electricity generation from renewable sources [17]. In this context, artificial intelligence (AI) methods (see Sect. 10) can significantly enhance modeling outcomes.

A crucial direction in modeling electric power systems and industrial processes is the creation of their digital twins. Digital twins are virtual replicas of physical objects or processes that enable energy-saving measures and improvements in the following technological and management processes:

- They allow for the modelling and optimization of production processes in a virtual environment before implementing changes in actual production, reducing risks and costs associated with experiments on real equipment;
- They facilitate predictive maintenance by forecasting potential failures and malfunctions. A digital twin can simulate various equipment operating scenarios and predict when maintenance or component replacement may be needed;
- They improve product design based on data from digital twins. Engineers can test different configurations and materials in a virtual environment, accelerating the transition to new product production and reducing costs associated with physical prototyping;
- They allow for employee training on new equipment or in new production conditions through virtual simulations;
- They optimize supply chains and logistics by modelling different scenarios and selecting the most efficient solutions.

The integration of these components provides a foundation for effective digital transformation in industrial enterprises. However, it is essential to remember that

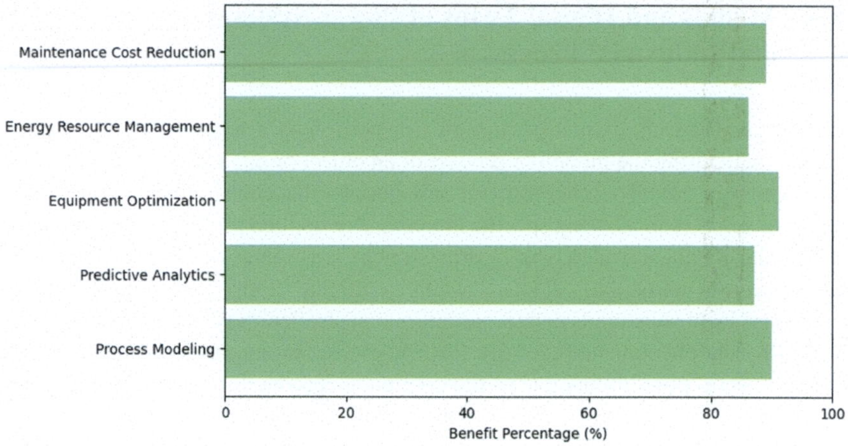

Fig. 11.4 Impact of digital twins on electricity savings

successful digitalization requires not only technological changes but also a transformation of corporate culture, the development of digital skills among employees, and a review of business processes.

Additionally, cybersecurity must be taken into account, as an increase in connected devices and data volume creates new vulnerabilities, as illustrated previously with digital substations and smart grids. Enterprises should invest in robust protection systems and employee training on information security fundamentals.

It is also important to note that digitalization is a continuous process. Technologies continue to evolve, and enterprises must be prepared for the ongoing updating and adaptation of their systems and processes. This requires fostering a culture of innovation and continuous learning at all organizational levels. The importance of HR issues and decisions was discussed in detail in Chap. 1.

Figure 11.4 illustrates the impact of digital twins on electricity savings.

The results shown in Fig. 11.4 are achieved, as noted above, through real-time process modelling using digital twins. Through predictive analytics, digital twins enable the forecasting of potential equipment issues and the avoidance of downtime, which not only reduces energy costs but also increases overall production efficiency [18].

Additionally, the use of digital twins helps lower equipment maintenance costs, as predictive analytics can determine in advance when repairs or servicing are needed, preventing emergency situations and associated energy costs.

Control Questions and Tasks

1. List the new technologies from the "digital transition" toolkit that improve energy efficiency in industrial enterprises.
2. Identify: a) IoT devices and b) activities they support to enhance energy efficiency.
3. What is "predictive maintenance" for equipment? How does its application benefit an industrial enterprise?

4. What are digital twins? How are they used, and what benefits do they offer to industrial enterprises?
5. Describe basic methods (mechanisms) to improve energy efficiency in industrial enterprises.

References

1. Mitrofanova IV (2020) Digitalization of industry as a factor in improving labor productivity and product quality. J Econ Sci 6:45–53
2. Gerbert P, Lorenz M, Rüßmann M et al (2015) Industry 4.0: the future of productivity and growth in manufacturing industries
3. Parviainen P, Tihinen M, Kääriäinen J, Teppola S (2017) Tackling the digitalization challenge: how to benefit from digitalization in practice. Int J Inf Syst Proj Manag 5(1):63–77
4. Suvorov AV (2023) Industrial enterprises in the digitalization era: tasks, challenges, and opportunities. J Indus Econ 19(3):112–119
5. Kagermann H, Wahlster W, Helbig J (2013) Recommendations for implementing the strategic initiative INDUSTRIE 4.0: Final Report of the Industrie 4.0 Working Group
6. Klimova NI (2021) Economic effects of digitalization at an industrial enterprise. J Inf Technol Econ 3:30–42
7. Henning S, Hasselbring W, Burmester H, Möbius A, Wojcieszak M (2021) Goals and measures for analyzing power consumption data in manufacturing enterprises. J Data Inf Manag 3(2):65–82
8. Saha N, Aulia MM, Rahman MM & Khan MS (2024) IoT-driven cloud-based energy and environment monitoring system for manufacturing industry
9. Mirani AA, Awasthi A, O'Mahony N, Walsh J (2024) Industrial IoT-based energy monitoring system: using data processing at edge. IoT 5(4):608–633
10. Schuh G, Anderl R, Gausemeier J, Hompel MT, Wahlster W (2017) Industry 4.0 maturity index: managing the digital transformation of companies
11. Parida V, Sjödin D, Reim W (2019) Reviewing the European industry 4.0: benefits and challenges of digital transformation in industrial enterprises. Technol Forecast Soc Chang 146:119–130
12. Kuzenov VV, Ryzhkov SV, Varaksin AY (2024) Development of a method for solving elliptic differential equations based on a nonlinear compact-polynomial scheme. J Comput Appl Math 451:116098
13. Varaksin AY, Ryzhkov SV (2023) Mathematical modeling of gas-solid two-phase flows: problems, achievements and perspectives (a review). Mathematics 11:3290
14. Zhang L, Zou L, Sun Y et al (2024) A preliminary assessment of the usability of magnetoplasma compressors in scientific and technical applications. Symmetry 16:1200
15. Kuzenov VV, Ryzhkov SV, Varaksin AYu (2023) Simulation of parameters of plasma dynamics of a magneto plasma compressor. Appl Sci 13:5538
16. Choi EH, Ryzhkov SV (2023) Magneto-inertial fusion and powerful plasma installations (a review). Appl Sci 13:6658
17. Wang Z, Bu S, Wen J, Huang C (2025) A comprehensive review on uncertainty modeling methods in modern power systems. Int J Electr Power Energy Syst 166:110534
18. Dalenogare LS, Benitez GB, Ayala NF, Frank AG (2018) The expected contribution of industry 4.0 technologies for industrial performance. Int J Prod Econ 204:383–394

Conclusion

The radical transformation of the electric power industry, conceptualized as "Smart Grids" (sometimes referred to as the "Digital Transition"), has begun to be implemented in many developed countries. This transformation demands coordinated efforts from government agencies, private firms, utilities, scientific and educational institutions, and other stakeholders to address the numerous challenges associated with the adoption of this advanced technology.

Stakeholders recognize that the development of new components for electric power systems, such as smart meters, monitoring and geolocation equipment, advanced recovery systems, energy storage systems, and more, must be supported by the creation of necessary communication infrastructure, the modernization and updating of existing network components, the establishment of interoperability and integration standards, and the training of a qualified workforce. For this purpose, the IEC 61850 standard has been developed and implemented, which considers the digital substation as a critical element of the smart grids of the future.

As with any advanced technology, numerous tasks must be addressed. Among the most important are:

1. Expanded use of digital information and control technologies to enhance the reliability, security, and efficiency of the power grid.
2. Optimization of network operations and resources required to ensure cybersecurity.
3. Increased deployment of distributed generation, including unconventional renewable energy sources.
4. Development and broader application of demand management technologies to improve energy efficiency.
5. Implementation of smart technologies (automated, real-time interactive technologies that optimize the physical operation of devices and consumer equipment) for metering, communication of operational and network status, and automation of electricity distribution processes.
6. Integration of smart devices and consumer equipment.

V. Y. Ushakov et al., *Digitalization of Electrical Power Engineering*, Power Systems, https://doi.org/10.1007/978-3-031-95705-5

7. Implementation and integration of advanced energy storage technologies and peak load management, including Plug-in Hybrid Electric Vehicles (PHEVs) and thermal energy storage for air conditioning systems.
8. Providing consumers with timely information and access to participate in grid management.
9. Development of standards for the communication and interaction of network-connected devices and equipment, including the infrastructure supporting the network.
10. Identification and reduction of unreasonable and unnecessary barriers to the adoption of smart grid technologies, practices, and services.

The authors hope that this textbook addresses the tasks and tools required implementing the "Smart Grid" concept in sufficient depth for students and graduate students specializing in "Electric Power Engineering" and related fields.

The manufacturer's authorised representative in the EU is Springer Nature Customer Service Centre GmbH, Europaplatz 3, 69115 Heidelberg, Germany. If you have any concerns regarding our products, please contact ProductSafety@springernature.com

Printed and bound by CPI Group (UK) Ltd, Croydon, CR0 4YY
05/10/2026
02228157-0007